Implementing ITSM

From Silos to Services—Transforming
the IT Organization to an IT Service
Management Valued Partner

Randy A. Steinberg

Implementing ITSM

From Silos to Services—Transforming the IT Organization to an IT Service Management Valued Partner

Order this book online at www.trafford.com
or email orders@trafford.com

Most Trafford titles are also available at major online book retailers.

Printed in the United States of America.

ISBN: 978-1-4907-1958-0 (sc)
ISBN: 978-1-4907-3512-2 (e)

Trafford rev. 05/05/2014

 www.trafford.com

North America & international
toll-free: 1 888 232 4444 (USA & Canada)
fax: 812 355 4082

Other books by Randy A. Steinberg:

Measuring ITSM
Measuring, Reporting, and Modeling the IT Service Management Metrics that Matter Most
to IT Senior Executives
Trafford Press ISBN: 978-1-4907-1945-0

Servicing ITSM
A Handbook of IT Services for Service Managers and IT Support Practitioners
Trafford Press ISBN: 978-1-4907-1956-6

Architecting ITSM
A Reference of Configuration Items and Building Blocks For a Comprehensive IT Service
Management Infrastructure
Trafford Press ISBN: 978-1-4907-1957-3

The above books supplement this book by providing much more explicit detail around IT Service Management support services, metrics, architecture, organization and tooling functions.

In addition, readers may wish to contact the author at RandyASteinberg@gmail.com to get complimentary access to ITSMLib™ which is a comprehensive library of ITSM artifacts, templates, tools, white papers, process guides, work plans and other guidance that compliments the topics presented in this book.

Rethinking how IT operates in the 21st Century . . .

Management of IT is undergoing a major change.

The role of IT has now changed from engineer-to-order to service integration. What IT traditionally engineered, built, owned and operated can now be bought from many sources more easily in the marketplace. This can be done without inheriting the specific costs and risks of owning, supporting, building and managing an operating infrastructure. The traditional IT operating model of delivering IT to the business in the form of capabilities and assets (commonly referred to as technology silos) no longer works in an age of cloud computing, on-demand services, virtualization, outsourcing and rapidly changing business delivery strategies.

The IT executive that cannot clearly articulate the services they deliver, the IT spend for those services, and how those services are consumed by the business is essentially running a commodity operation that will be forever whip-sawed by business demand and will never been viewed as nothing more than an overhead cost to the business. In addition, IT organizations that continue to operate this way will see their budgets and services eroded away over time as business customers drift towards outside providers that operate with service-driven approaches and practices.

Pushing this further are technology vendors with new forms of IT Service Management (ITSM) tools and solutions. The latest generation of tools cannot be implemented successfully without an understanding of the services to be delivered, how information flows and how it is communicated throughout the IT support organization. This information is now baked into how technologies are customized to work effectively creating efficiencies throughout the support organization.

That IT organization itself is also changing with a major transformation away from leadership by engineering to leadership by service. Accountability for end-to-end service is becoming the new norm. Gone are finger pointing across technology silos that used to place service integration activities in the hands of IT executives. This new accountability is now being pushed down lower into the organization and helping middle managers cope with this change is one of the key challenges to overcome.

Many IT organizations still have a long leap to take in this transformation. It is the goal of this book to provide specific guidance to IT organizations undertaking this effort. That guidance is based on many years of front line experience helping many IT organizations around the world effect this transformation.

It's time to operate IT like a Service Organization—happy journey!

—The Author

Dedication

This book is dedicated to those very hard working IT professionals and managers who deserve to see their IT solutions deploy and operate day-to-day at acceptable costs and risks to the business organizations that they support.

Table of Contents

Chapter

1

Introduction

> "Any investment in an IT solution that cannot be deployed and operated
> day-to-day at acceptable cost is an investment that is totally wasted."
> —Randy A. Steinberg

This book is unique.

It shouldn't be. Yet, having tracked around the globe helping many companies with their IT Service Management (ITSM) issues I find it amazing that such little information is available on how to actually use ITSM concepts and practices to support and deliver IT services. While much information exists about ITSM certification, training and how the processes work, almost no one has addressed how to use ITSM to transform the IT organization from disconnected silos of technology capabilities to a valued service partner.

Until now.

IT can no longer operate and manage itself the way that it historically has in the past. In the good old days, IT was managed like a back office operation as independent technical silos that didn't always work well together. There was little need to have to deal with end-users and customers. Guess what? In today's world, IT is managed like a back office operation as independent technical silos that still don't always work well together. There is still little effort on dealing with end-users and customers as evidenced by the dearth of service quality interactions with those groups. How many have a Service Desk located in the United States?

What's different you might ask?

IT was formerly driven through an engineering focus to design and build IT solutions. In today's world, what with Cloud Computing, virtualized systems, packaged software, bring your own device or "apps", much of what took past efforts to design and build can now be bought on the open market. The role of IT has now changed to a focus on service integration. IT organizations that fail to recognize this fundamental shift will soon be left far behind.

The combination of increasing speed of technology changes, the global business economy and pervasiveness of computing have now placed more demands on IT that ever before. IT must now manage itself by the services it delivers versus the technologies it uses. If organizations fail to adapt, they will see management costs continue to go up, service quality to go down and complexity to increase.

In today's world, computing is now so pervasive that your company's customers, employees and your most vital business functions are exposed directly to how IT services are managed and operated. Try getting along without any E-Mail availability. Try doing online banking on a Sunday morning when the IT department has shut off its servers for maintenance activities. Ever announce a new product with guaranteed shipping in 24 hours when your IT infrastructure can only operate with a five day processing window?

Something funny has happened with IT over these last many years. People got so excited about new technologies and ways for doing business that they forgot to look at the ongoing costs of maintaining and running their technology solutions. Think this doesn't happen? Here are some examples from today's headlines:

A major fast food organization invests millions of dollars in an effort to put all new automation technologies into its restaurants. About 2 years later and $300 million dollars poorer, the company discovers that the costs for maintaining this new technology are too prohibitive.

A large federal agency signs a major deal with a popular job employment web site vendor to speed up job postings and applications for federal agencies. After implementing this solution it turns out that the web site cannot handle the capacity of business volumes experienced by the agency. The last quote by a Senior Federal Official: ". . . we're back to the drawing board on this one . . ."

A major retailer completed a merger agreement with another large retail chain. Seems someone forgot to put two-and-two together to realize that one of the retailers would no longer need the services of their newly hired IT outsourcing provider. Cost of delay in notification to that provider? About $48 million.

A major inner city school district happily receives 5,000 laptop devices donated by a local corporation. Within 3 months about a third of the machines are stolen, lawsuits are put against the school for inappropriate content found on two of the machines and it is later found that the rest cannot be economically hooked up to the school intranet.

A large banking concern implemented a new online banking system for their customers. Two years later, a new CEO was surprised to discover that the ongoing management of this system was costing the bank over $3,000 per bank account per month!

A major shoe manufacturer implements a new order processing system in time for the Christmas season and soon discovers that slow system performance is creating a major backlog of orders to be processed. The company can only resolve this issue by forcing employees to work overtime and adding significant numbers of temporary administrative staff to do nothing but enter orders. This got worse when someone remembered that a slew of additional infrastructure had to be temporarily procured so the additional staff could enter the orders.

Lastly, who can forget about a national major healthcare website critical to providing healthcare coverage for millions of citizens that appeared to be poorly prepared to handle transaction volumes, was disconnected to most of the back end processing systems and caused major political issues and delays? Even worse, the public officials' response: "Everyone knows IT solutions never work when they are first rolled out . . ."

It is headlines like those that provide evidence of the massive amount of waste in IT investment that goes on. It doesn't take much to get to the billions of dollars every year for business organizations. These are unbelievably expensive mistakes. In addition, they typical IT budget is heavily skewed (70% or more in most cases) towards just keeping the IT lights on. Very little is left for innovation. Is it no wonder that businesses are looking for ways to drive down the costs of IT?

Despite all this, information on how to run and operate an IT infrastructure is surprisingly lacking. Bookstores are far from teaming with IT management literature. College courses in IT management are few and far between (and UNIX Administration 101 does not count here). Knowledge and experience are highly dependent on individuals who may or may not have the appropriate expertise. How can something so vital to the IT community receive such little attention?

You get the idea. Much notice is typically given to an exciting new technology or business initiative and short shrift is given to their deployment and operation until it is too late. This leads to one of the primary reasons for employing an IT Service Management culture and operation:

Protect corporate investments in IT solutions and infrastructure by ensuring that they will be deployable and operable on a day-to-day basis at acceptable cost to the company.

Another argument you sometimes hear from IT itself might go like this:

We've done a wonderful job for years! There's no reason to change! We always place service and the customer at the highest priority in our organization!

Dig deeper and you find that Customer Satisfaction ratings, if they exist at all, are done infrequently and not tied to any actionable service improvement program. Worse yet, they may have an IT support organization structure that resembles a configuration diagram. You probably recognize this. It shows an IT organization with a midrange operations group, a network group, a desktop group, a legacy group, database group, security group, web infrastructure, etc. Perhaps the organization even names those groups after technologies like the e-cloud group, network operations or the server department.

This kind of organization, which is pretty typical in most companies, usually leads to this kind of problem:

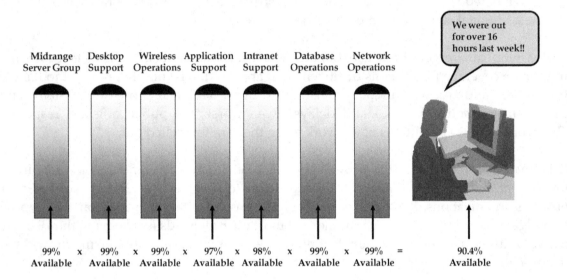

Perhaps you've seen this before. Each IT unit delivers impeccably, but the end user or customer still complains. *There is no accountability in the organization for the actual service received by the end-user or customer.* In many organizations you actually have to get to the CIO or Senior Management levels where these finally tie together. No wonder business units get frustrated. Some of them may even implement their own IT departments who replicate services that should have been done by the corporate IT function in the first place.

What's really needed is something like this:

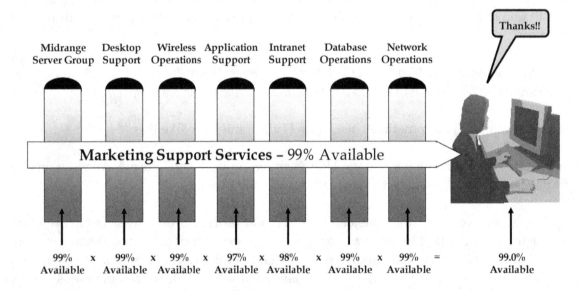

Is the source for all this that IT people don't know any better? I think not. I find IT people are pretty smart in general, but what they are lacking is a system for how to support end users and customers effectively, manage expectations and, in general, align IT work and priorities to the business.

This is where IT Service Management comes in. It represents a major shift from providing IT services in a hundreds of parts and pieces to a seamlessly aligned service organization where all those parts and pieces work fluidly together to meet business objectives. That is the essence of an IT Service Management solution.

It's not about each individual process.

It's about how those processes work together and interact to provide services aligned with the business.

Yet, disappointingly, much of the focus on ITSM efforts to date has been on individual processes. Many companies attempting to implement ITSM will typically implement process by process. Experience shows that many of these efforts will end up re-architecting processes later on as other processes come into play.

When speaking at various ITSM related gatherings, I usually take a quick informal poll with the audience:

How many are implementing Incident or Problem Management?
(Lots of hands go up)

How many are implementing Availability Management?
(Usually zero or 1-2 hands)

How many are implementing IT Financial Management?
(Usually no hands)

How many are aware that he top two issues facing CEOs/CIOs for the last 5 years has been the costs of managing technology and providing availability of services?
(Silence from the audience)

Talk about an IT/business alignment gap! The point is that this is about implementing IT Service management, not individualized processes. That is why the approach taken in this book advocates first understanding the business issues you are trying to fix, then implementing the pieces and parts of ALL processes that address them at the same time.

Having said this, IT people will then ask: "Whoa! Too much change at once! How is this possible?" A good question, but it demonstrates the silo mentality is still alive and kicking. Who said that all processes need to be implemented completely? Could you not pick the parts and pieces out of each process that provide the most benefits first and then phase in others over time? Could not those pieces be implemented within the context of an overall ITSM strategy? This book will show you some very practical approaches for prioritizing your activities to get the biggest bang for your efforts across all processes.

Here are several last thoughts on typical questions from IT:

Which process do you start with first?

Answer: The ones that best address your current IT business issues. One can make a solid argument for how any one of the processes needs to be done before the other. This just illustrates why all should be addressed simultaneously. Each ITSM process has many dependencies that with all the other processes. One look at this and it is easy to see how each ITSM process *interlocks* with all the others.

Which is more important: Process or Technology?

Answer: Process every time. A very senior director of technology once challenged me on this. He basically felt that process work is not needed—this can all be encapsulated into technology. To which one might ask: "Where do you get your design specs?"

Think about that a minute. Not one piece of operational technology could ever be designed and built without some understanding of the processes and workflows that need to be accomplished. All technology serves processes. I would even bet that many IT people select and purchase products hoping to gain a process by doing so.

Are not vendors building smarter tools that will eliminate the need for process?

Answer: There are some promising technologies on the horizon in the way of self-healing and autonomic technologies. Surprisingly, these tools will actually increase the need to understand what services and service requirements need to be in place. This knowledge will be necessary to customize these tools to obtain the promised benefits. Think tool vendors do not recognize this? Almost every major tool vendor has been building ITSM expertise with their support staffs over the last few years. In addition, newer IT Service Management tools literally require that you fully understand your services, support teams and process workflows or you will have a disaster on your hands.

At this point it's now time to step down off the soap box and get practical. The remaining chapters are about how to work with IT Service Management. Think of ITSM as the overall set of practices and methods by which you will design, build and operate your services.

The core subject of this book is to provide an overall strategy and framework for implementing ITSM in 6 to 9 month waves. If you were looking for a better way to do Service Level Management or Problem Management (for example), this is not the place. Nor is it a beginner's manual for ITSM processes and concepts. It is already assumed that readers already have some basic level of familiarity with ITSM fundamentals.

The following presents a quick synopsis of each remaining Chapter:

Chapter 2—Nine Things That Work

We'll start off with an examination of what common things successful companies have done to implement ITSM solutions. These provide guidance and input for how the suggested implementation approach has been structured and organized.

Chapter 3—ITSM Transformation Overview

In this chapter, we'll cover the ITSM transformation approach at a high level. This will give you a good feel for how everything comes together.

Chapter 4—Adapting to an ITSM Culture

Think you can design and build your ITSM solutions, toss them out to the organization and then expect everyone to use them? ITSM solutions require a successful organizational change effort. This chapter reviews some basic organizational change concepts and describes some practical ways to develop an ITSM Communications Strategy and perform Stakeholder Analysis. Just about every company who has undergone an ITSM initiative has cited organization change as absolutely critical to success. That's why this chapter has been included.

Chapter 5—Transformation Organization

How should you organize your ITSM transformation team? What roles and activities are needed? How should your company organize to execute ITSM solutions and activities? Find answers in this chapter.

Chapter 6—ITSM Visioning Work Stage

This is the first Work Stage in the ITSM Transformation effort. Find out how to establish a Service Vision, Program Success Factors and prioritize what aspects of a full ITSM solution are the most important to your company.

Chapter 7—ITSM Assessment Work Stage

Learn about conducting holistic assessments that cover technology and organization in addition to processes. Find out what gaps exist between the current state of your IT organization and the planned Service Vision. Identify and prioritize actions to close those gaps. Identify Initial Win Projects that will gain noticeable benefits early in your implementation efforts.

Chapter 8—ITSM Planning Work Stage

Learn what is needed to effectively plan and organize your efforts to ensure a smooth running implementation program. A must-read if you have been put in charge of a large transformation project.

Chapter 9—ITSM Design Work Stage

Learn a step-by-step approach towards designing your desired ITSM solutions.

Chapter 10—ITSM Initial Wins Work Stage

This chapter presents a recommended standard project lifecycle for each Initial Win you will be implementing. The chapter also includes an inventory of common Initial Win Projects that other companies have found valuable.

Chapter 11—ITSM Transition Work Stage

In this chapter, you will find out how to report and govern your implemented ITSM solutions. Learn how to "keep the momentum going" for the next wave of solutions. Learn how to transition and organize to operate ITSM solutions.

Chapter 12—ITSM Tooling Considerations

Presented here is a set of generic tool requirements from each type of tool needed to support your ITSM processes.

Chapter 13—ITSM Process Work Products

Use this chapter as a reference for all the key artifacts needed to be produced as part of a comprehensive ITSM transformation initiative.

Chapter 14—ITSM Operating Roles

In this chapter are listings of each organizational role needed to operate each ITSM process or function once in place. It also includes descriptions for these roles to help you build your IT Service Management organization and matching job descriptions.

Chapter 15—ITSM Design Principles

Presented here is a starter set of Guiding Design Principles for each ITSM process solution and function.

Chapter 16—Continual Service Improvement

Presented here is a suggested ongoing programmatic approach for identifying service improvements on an ongoing basis and executing on improvement initiatives.

Chapter 17—Building a Service Step-By-Step

Presented here is a suggested approach for building a new service from the ground up using many of the concepts and methods discussed in this book.

ITSMLib™—A Comprehensive Service Management Program Aid

As part of your efforts, and with this book, you have free access to one of the largest ITSM knowledge portals available. ITSMLib provides access to real world working documentation, templates, tools and examples for almost any ITSM project. The library is structured to easily find knowledge and can be easily searched with phrases and keywords to find relevant information. We carefully screen items to eliminate sales pitches and proprietary content such that just about everything here comes from actual field use. As a result, the artifacts, tools and templates are not being touted as "best practice", or may not always look professional, but you may find them useful in jump starting your own solutions with ideas and content that has worked for others.

The ITSMLib mission is to substantially increase knowledge and skills in IT Service Management, deploying and operating IT solutions with acceptable costs and risks. This is an area that has sorely been neglected by the IT industry in the past incurring great costs and service failures. It is hoped that ITSMLib will in some small way turn the corner on what is a major industry problem significantly boosting every IT practitioners' ITSM skills and capabilities with real-world sharing of solutions as they are being implemented across the industry. More importantly, much of the concepts discussed in this book can be found there with actual artifacts, templates and methods used in real IT organizations. The library is continually expanding.

Access is on an individual basis only. Requests for access can be made to:

RandyASteinberg@gmail.com

It is very much hoped that you will find this book very helpful. I think you will find it truly unique. It is devoted to providing you with a path and direction for getting your IT Service Management solutions implemented in a logical and orderly manner. Hopefully there will be much fewer questions in the future like:

"How do you actually work with ITSM?"

If even one company is helped by this book, the effort for writing it will have been worthwhile.

Chapter

2

Nine Things That Work

"CIOs need to understand the relevance of service management as a component of IT. Fundamentally, it stops things breaking and makes them work more effectively."
—Kerry Crompton, CIO

A company cannot enhance their IT service quality by throwing money at hardware and software.

Before starting an ITSM transformation effort, it helps to understand what has worked for others who have gone through this before. There are a number of lessons learned that seem to be common across many organizations that have been implementing Service Management solutions. I have summarized this into nine key behaviors that have appeared again and again in those efforts that achieved significant results. They are:

1) Treating the effort as an Organizational Change effort versus an IT Project

2) Creating a balance between strategic efforts and initial wins

3) Implementing ITSM As An Operating Culture—Not Just Selected Processes

4) Targeting 20% of the effort to get 80% of the benefits

5) Staffing the effort with a balance between leaders and managers

6) Establishing a compelling business reason for the effort

7) Scoping the effort based on service delivery versus location or business unit

8) Getting Senior Management to "walk the talk" as well as sponsor the effort

9) Recognizing the critical importance of metrics

Now let's take a more detailed look at each of these on the following pages.

#1—Treat the Effort as an Organizational Change Effort

Make no mistake about it. An ITSM transformation is first and foremost an Organizational Change Management effort. It is not an IT project.

Many companies fall into the trap of starting up an ITSM transformation effort by assigning it solely within the IT organization. The effort gets treated as another IT project. Staff is selected from within IT sometimes with little or no experience beyond technical or process implementation skills. Many of these attempts soon stall or fail.

In reality, a successful transformation may result in many changes to how people go about their work. It will have impact and changes not only on the IT staff, but on many non-IT parts of the business as well. It will require winning the hearts and minds of the organization to adopt and ingrain new techniques and ways of working. There may be many other quality, business and IT initiatives already in flight that might be leveraged for greater success.

How do you go about uniting all of these together? If the current IT organization is organized around technological pillars, what changes, organization and behavior, will need to be adopted to get them to work together seamlessly to deliver a service? How will you deal with the many factions of the organization that may or may not be on board for this effort? How will others outside of your project be convinced about the importance of ITSM? Who are your customers and stakeholders? How will you manage those stakeholders? How will you keep momentum going throughout the effort?

Organizational Change Management is a discipline that addresses these kinds of issues. It is no small task. The amount of effort to strategize and execute a strong organizational change strategy is typically underestimated by most people. This may be surprising, but most companies that have begun implementation efforts are quickly finding that well over 70% of the implementation effort is spent on organizational communication and change activities.

Defining and building ITSM solutions by themselves will not work without an organizational change effort. This one element was cited by nearly every organization that had experienced an ITSM success. Without it, there is great risk that the effort itself may slow or die before completion. Worse yet, the effort may complete with the end result being shelf ware on someone's desk.

The trail of ITSM process projects is littered with transformations that ground to a halt after several months, ended up with vastly reduced scope or simply failed to live up to the business benefits that everyone hoped to achieve. The organizational change aspects of this effort need to be the centerpiece of the implementation approach. Although you might feel initially uncomfortable with this, you will be greatly surprised at what a difference this will make in achieving success.

#2—Balance Strategic Efforts with Initial Wins

It is surprising how many times you hear the following statement from organizations planning an ITSM transformation. It usually goes something like this:

> "This will be a major effort for the IT department. We expect that this will be a 2-3 year effort. This is only the first step in a long journey".

It would be an interesting task to visit these organizations in 2-3 years to see if there is even an ITSM project still running. In today's world, very few businesses have the patience and funding for projects that won't show benefit for several years. It is most likely that transformation projects scoped in this manner will wither and die on the vine or be abandoned for other priorities.

In today's business climate, benefits must be shown within a year or less. Probably within the first 6 months. Yet, ITSM is large. It has many processes, key activities, and changes that must be made. It has new organization roles and responsibilities, maybe even new technologies that must be implemented. How can this be done?

The answer lies in balancing short term gains with the more strategic gains. A fully redesigned process grounded with the ITSM best practices might include new organizational roles, new responsibilities, metrics, reporting and new technologies. This will most certainly take longer to implement than 6 months. Is there a subset of this grand solution that can be put into place much more quickly? Can a short term solution be rapidly put into place that addresses an acute business need? Is it possible to target only the top 5, 10 or 20% of the solution that will meet 60-80% of the need?

Identification of short term gains has many benefits. One is that they are usually specific enough that most company staff can relate to. They may not yet understand how and why reaching a higher process maturity in Incident Management (for example) will work, but that short term effort to have a common classification scheme for logging incidents is certainly understandable. Not sure how Configuration Management will ever operate, but we can certainly gather and collate all IT documentation, designs and policies, throw them in a file cabinet and manage the file cabinet like a configuration database. Ideas like these are pretty specific and understood by implementation staff. They can easily see where the implementation effort is going and gain a sense of accomplishment which will boost acceptance and cooperation in getting things done.

Another benefit is that the business will see something that can be touched and felt. This will serve to excite people which in turn will build a stronger coalition to go onto each next implementation phase. Lastly, it will allow the implementation team to gain some immediate wins early on in the effort as well as positive recognition from the business.
While there are many benefits from the short term gain approach, the trap to watch out for is to not lose sight of the strategic goals. After all, this is why the overall effort got started in the first place. Focus too much on the short term gains and the strategic benefits may get lost or never realized. Focus too much on the strategic side and process designs soon

become shelf ware. Therefore, it is important to make sure both are being addressed in the transformation effort with enough balance to ensure that short term gains are guided by ITSM principles and serve as stepping stones to the more strategic solutions.

#3—Implement ITSM as an Operating Culture—Not Just Selected Processes

Every ITSM process has critical dependencies upon another. You cannot implement Configuration Management without Change Management to control changes to the Configuration Management Database. You cannot perform many Service and Business Capacity planning tasks unless service levels and a Service Catalog have been built. Problem Management cannot provide much benefit unless incidents have been recognized and properly classified by Incident Management. Critical strategies operated by IT Service Continuity Management cannot be properly considered without an understanding of financial and availability concerns from Financial Management and Availability Management. And so on and so on.

No ITSM process is an island unto itself. They are like a beehive in that any one of the ten processes underpins the other nine. Yet for some reason, most IT organizations will limit their scope to only a few. When asked why, the usual reason given is concerns over the amount of change that can be absorbed at one time by the business or just not enough resources available to handle the implementation workload. This gets back to the common IT problem of thinking in pillars instead of solutions.

When looking back at many transformations that start this way, it is found that many of the perceived benefits of the processes undertaken are never achieved. Transformation teams leave out key pieces of their solutions citing things like ". . . that'll be done later when we get to Change Management". Some of those process efforts may even stall when the team suddenly is blocked from proceeding further until an output from an unplanned process is in place. They may implement processes with workarounds or other means that then are incompatible with later processes requiring replacement or reengineering before those other processes can be put into place.

The worse crime is that the wrong processes may have been chosen to be tackled first. As an example, imagine all the effort that a company spends on Incident Management when what was really needed was Availability or Problem Management. This actually occurred in one case. Senior Management wanted the pain of severe service outages to go away and instead ended up with the outages still occurring—but at least IT did a better job of reporting and logging them!

When asking IT organizations why they do this, one of the key reasons cited is a mistaken notion that taking on many ITSM processes at once will create more change than the business would be able to handle. Unconsciously, they were thinking that at all of them must be completely designed, built tested and implemented to their fullest extent as described in the ITSM literature. This does not have to be the case. There is nothing wrong with implementing some processes fully and others only to a small extent. They key lies in understanding what actions will benefit the business most and then focusing your efforts within the context of an overall ITSM implementation program—not a process project!

This is why the recommended transformation approach described in this book starts with an ITSM Strategy effort. That effort will set the stage to guide you on where to place your program priorities. Based on those priorities, you will determine what is needed for short term gains. You will determine which processes may have to be built out completely versus just enough to satisfy some key pain points. For those process not fully implemented, further work might be handled in subsequent ITSM transformation efforts.

By addressing all of the processes simultaneously you will be implementing an ITSM service culture instead of just part of it. You will also gain considerably more buy-in and recognition from a wider audience within the business organization. More importantly, you may uncover needs and priorities from other parts of the organization that will provide significant wins that were not foreseen previously.

A wonderful example of this was with a large financial services firm that took this approach. The IT department rightfully recognized that some core basics in Incident and Problem management were needed. Almost everyone on the team scoffed at Financial Management. "Not much needed there at this time . . ." was frequently heard. Early on in the implementation effort, when meeting with the Financial Management business stakeholders, it was discovered that much effort was taking place outside of the IT department to map the technical infrastructure to financial products produced by that company. The intent was to determine how much the company was spending on supporting these products and compare it to the revenue brought in for them. Low revenue producing products could be phased out by the business if it was determined that their support costs were high.

This turned out to be a significant deal for both IT and the business. Boosting both Configuration Management efforts and Financial Management efforts with the short term gain of understanding product revenues and support costs suddenly became critical and turned into a big win for the company. The business began to quickly realize benefits of the ITSM effort that could actually be measured on the bottom line. Senior Management suddenly saw what could be accomplished and a new set of converts were added willing to provide additional funding and support for ITSM initiatives.

#4—Target 20% of the Effort to Get 80% of the Benefits

A large credit card processing company had recently adopted ITSM hoping to improve the reliability and quality of their IT services. They sent a number of their IT staff people to ITSM training and established an ITSM competency within the IT department. The staff came back from training and was solidly convinced that no ITSM efforts could start unless a Configuration Management Database (CMDB) was in place. It is not clear where they derived this misconception, but they soon funded a project to begin the detailed design of a CMDB. Consultants were brought in and began the design work.

Since this was a large company, there were tens of thousands of Configuration Items (CIs) such as PCs, servers, networking devices and so on. Major amounts of time were spent on data design for how the CMDB should look. The project team worked hard to develop a database schema that would cover every conceivable CI type and relationship the company had. Every hardware type, software type, network type, application type and so on. The team worked over several weeks and came up with some pretty solid data designs for the CMDB. What do you think was the end result?

The end result was a waste of time and money. The entire effort completely stalled and soon died a quick death. The company never moved forward with anything ITSM related after that. What went wrong?

The designs done by the team were solid. It was good work considering the complexity and size of the organization. When the IT department looked at the finished design, the number of CI types, relationships and information needed to populate the database with were enormous. Information on over 50,000 PCs, 8,000 servers, untold numbers of routers, switches, 6,500 applications and much more needed to be input and validated into the new CMDB. Although there were asset repositories throughout the company, the task of linking these together and supplying the relationships between them was absolutely daunting. Nevertheless, a project plan was developed, tools were selected and the company management was told that with some good project management practices, a CMDB could be delivered within 9-10 months using a 12 person project team.

To a CIO battling to instill companywide quality standards, eliminate outages and align IT services to the business, the prospect of spending that amount of time and money made no sense. Although the team argued all the good merits of why a CMDB was needed, it all came to naught. Project killed, funding removed. Since everyone was under the mistaken belief that no ITSM effort could move forward without a CMDB, this meant that the entire ITSM initiative was over before it could get off the ground.

In looking back at the above, it never occurred to anyone that there might have been a simpler path to get the value and benefits out of a CMDB to start with. Did every PC model and serial number have to go into the CMDB right away? Did every component of hardware, software, network and applications have to be accounted for before anything else could start? In addition, it was noted that the focus of the design was pretty techno-centric. Where were the business services? The financial products the company sold? Why did CMDB design start at the lowest technical level of IT instead of at the top starting with the company's customers and services? Was there more value to having low level technical detail versus using a simpler approach?

IT people love to reinvent and tinker with new solutions. It's in their blood. As in the above example, many IT people love to reinvent solutions making them as perfect as possible. There are tendencies to focus on detail in the extreme. Resist these temptations wherever possible. Target the 20% of the effort to be taken that will satisfy 80% of the need. Whenever possible, utilize as much of what may already exist within the organization first. Design something new only when absolutely necessary. Falling into the over-design trap is fraught with great risk, project delays and project funds that could be better spent on solving a wider set of IT problems and issues.

#5—Balance Efforts with Good Leaders AND Managers

Much attention will be paid in this book as to how to properly staff up and organize for a successful ITSM transformation. This is not an effort to be taken lightly. Some organizations have attempted this as a part-time activity for the IT staff. Others have populated the effort with IT clerks, analysts or quality assurance staff that have little experience, an ax to grind, technology or other narrow focus. Some will use the project to further other non-related project or activity goals. Many companies will staff the entire effort solely through the IT operations department. All of these approaches have one thing in common. They fail.

It is important to understand the difference between those who can manage and those who can lead. Both types will be needed. A manager is someone who has the skills and attention to detail to execute on ideas and tasks that meet transformation objectives. A leader is someone who is willing to challenge the status quo, research new ideas, fire up the project team, break old habits and company boundaries.

Learn to recognize these traits and the differences between them. A manager placed in charge of an ITSM transformation effort will most likely try to solve problems and issues within the context of how the business operates today. They will typically not challenge the status quo and instead focus and limit solutions to the current company culture and organization. Solutions will typically be shoe-horned into the current organization structure and style. Major business benefits may be lacking or totally missed at the end of the transformation effort.

A leader placed in charge of an ITSM transformation effort will always challenge the status quo. They will generate radically new ideas and constantly question why things are done as they are today. They are not afraid to step outside the box, try new ideas and challenge the existing organization structure. However, there always lurks the danger that a leader may not be able to take the transformation effort down a path from start to finish. Good ideas with business value and benefit may end up half implemented or never come to completion at all.

You need both to properly balance the transformation effort. Make a concerted effort to look for both and place them in appropriate roles within the project.

Another area to consider is making sure that all of the business is represented on the project effort—not just the IT operations department. By leaving the project solely within IT operations, you limit the ideas and focus of the team. You may also miss the mark on achieving true business benefit. Worse yet, without this there is a significant risk that the overall effort may be suddenly stalled or killed at deployment time due to unforeseen business issues or requirements that were never considered.

The transformation organization structure is discussed at great detail in this book to help you specifically avoid the kinds of land mines that can derail a successful ITSM effort.

#6—Establish Compelling Business Reasons for ITSM

From the outset it is absolutely critical that compelling reasons be documented and articulated as to why the transformation effort should take place. It is not enough to simply qualify the effort as a way to introduce best practices for systems operational management or instill ITSM concepts in IT staff. There needs to be a compelling business reason (or set of reasons) for undertaking the effort. These reasons need to match the following criteria:

- They identify true pain that the organization is experiencing or about to experience

- They need to be addressed now or dire consequences will occur

If one or more stated reasons fit both of these criteria, then you will have a solid business case that will not only get the attention of Senior Management, but also keep the momentum going throughout the effort. An additional benefit will be that a much clearer focus and set of priorities will exist as transformation activities take place.

Examples of compelling business reasons might include:

- Need to stop the bleeding from major outages and availability problems

- Desire to reduce non-value labor in the IT organization

- Need to demonstrate best practices in order to differentiate the business in the marketplace

- Need to meet upcoming regulatory or audit criteria

- Business has shifted to a customer-centric delivery model and is looking for IT to do the same

- Need to reduce outage durations and general angst in the organization due to finger pointing and overall confusion when incidents occur

- Need to cut infrastructure costs yet maintain quality delivery of services

- Planned acquisition or merger will be put at risk unless management of the infrastructure is brought under control

- Need to eliminate management redundancies that exist throughout the IT organization

- Increase in technology complexity is making the infrastructure harder to manage using current practices and will soon result in more outages and higher delivery costs.

There may be other reasons unique to your organization as well. Bottom line: make sure there is a business problem to be solved and one that needs to be solved immediately.

There is also another trap to be made aware of. This concerns the common practice of using Capability Maturity Model (CMM) process maturity scores as justification for a transformation effort. Many consulting organizations jump into assessments early on and use these scores to help a business assess and analyze the IT processes within their organization. Developed from this work are score ratings that gauge the level of process maturity that may exist.

For example, a zero score might mean there is a complete lack of awareness within the organization about a particular process. A one might mean awareness exists but no process ownership or documentation. Higher scores imply more evidence of solid processes in place, automation and linkage to the business. The consulting world in particular has developed scoring and analytical approaches that not only score the maturity of the processes currently existing within an IT organization, but also provide a score comparison with other companies of similar size and industry.

This kind of an approach is somewhat attractive to Senior Management. IT, in general, tends to have challenges in presenting meaningful metrics and statistics that are usable by upper management. All of the sudden, management is suddenly presented with a process maturity score and can see how well that score stacks up against their competitors and peers. Upper management is then told that with the additional help of consultants and implementing ITSM, the maturity level can be raised to a higher standard that is equal or better than the peer groups they have been compared to.

Here is where the real danger lies. While there is some value in using this kind of approach to assess current state, company management may fall into the trap of viewing the maturity score as the end goal of the effort. There have been cases where an ITSM transformation was put into place with the goal stated in terms like ". . . we will raise the maturity level from 1.8 to 3.1 . . ."

Lost in all this are the real business benefits that are to be achieved. What does the company gain by "getting to 3.1"? Does every process need to be at "3.1"? If not, which processes? Will "3.1" actually fix our availability problems or reduce incidents and outages? The key danger here is this:

> If you use a process maturity score as an end state, you have only ASSUMED that the business problem Senior Management wants solved will actually go away.

Having said this however, there is a place for these kinds of maturity assessments within an ITSM transformation effort. They provide a wonderful benchmark that can be used to gauge whether problems may exist inside your processes. A low score may indicate areas to focus on to find root cause within your processes that may be contributing towards a business problem. A later chapter in this book describes the ITSM Assessment Work Stage and presents ways for using maturity scores and ratings as a way to benchmark your ITSM processes appropriately.

#7—Scope Efforts by Service—Not by Geography

IT people tend to think in terms of hardware and software. You obtain a new hardware or software product and then you install it somewhere. Upper management may also share this view—"the IT department's job is to install things". Therefore, it is not unnatural to view an ITSM transformation like a hardware or software implementation effort. "After all", says management, "this is an IT project and that is what IT does".

Therefore, IT culture in hand, it is decided that the ITSM transformation effort for the first year (for example) will only cover a specific data center location or geographic region.

I once heard someone use the term 'sheep dip' to describe this approach. A company with many IT processing sites will scope the transformation effort to "sheep dip" one processing location with ITSM. Following this, another location will be "sheep dipped" in ITSM and then later on another and so on.

The problem with this approach is that it only works for those companies where each location is highly independent, both in IT function as well as business function. The case where this is true may be rare.

What's more likely is that a global organization may have products and services that cross many processing sites and locations. In addition, the customers being served by that business may also be global and expect consistency in service no matter where in the world they are using the services. Therefore, addressing only one processing location will provide extremely little benefit. One part of the IT organization may successfully execute ITSM, but overall service will still suffer the same problems as before since these poor practices still exist at the other locations.

Therefore, the correct solution is to ensure that ITSM best practices are put into place at EVERY point where services are being delivered. In a global organization, this means instilling these best practices around the world in all locations at once. While this may seem daunting at first, there is a way to organize the effort to make this happen much sooner than you think. A later chapter discusses organizational considerations and an approach for making this happen using the concept of a Coalition Team.

There may also be a high level of discomfort within the IT organization in attempting to work with other IT organizations from other locations. Despite these fears, you cannot avoid the fact that successfully implementing ITSM solutions in only one location will result in time and effort being spent, but little to show for it from a Senior Management and business perspective.

One other danger point to consider is transforming an IT sub-unit without an overall enterprise view. There have been companies that have done this. For example, a transformation effort is being done in the network group, another one in the server group and yet another effort started in the mainframe host group. Without an overall enterprise effort, business units soon become buried in things like multiple service catalogs, SLAs, competing OLAs, tool solutions, etc. Not a good situation and one where the business may question why ITSM was done in the first place.

#8—Get Senior Management to Walk the Talk

It goes without saying that Senior Management sponsorship and buy-in is needed from the start. No exceptions. Without this, the effort is literally doomed. While heroic efforts can be made without it by lower level staff, and even with some positive results obtained, experience has shown these attempts will ultimately lost steam in the long run. Without Senior Management buy in, attention and priorities will be placed on other efforts and the appropriate level of organizational change will never be reached.

Remember the saying:

> *Implementing ITSM without Senior Management sponsorship is like garden that initially blooms but soon gets overcome by the desert.*

Make sure that appropriate levels of management sponsorship are in place. Nodding in agreement that ITSM is a good thing and that more initiatives should be seen with it is nice, but not good enough. Senior Management needs to exhibit two critical behaviors:

- They are actually funding the effort and providing staffing resources

- They are "Walking The Talk" when it comes to IT Service Management

Without both of these, the effort will fail. Funding and staffing the effort are obviously important to make sure things get done. "Walking the Talk" ensures that all employees see that management is serious about making changes for ITSM. Without this latter behavior, employees may feel management is not serious or this may be "another flavor of the month"—one of many quality initiatives that seem to come and go.

There is an interesting example of this with a large Credit Company. Senior Management was all for the ITSM initiative. Funding was put into place, staffing resources were provided. Despite all the upper management buy-in, the worker bees decided management wasn't serious enough about the effort and did not place it as a priority. As a result, the effort failed within its first 6 months.

What does it mean to have Senior Management "Walk the Talk"? It means that senior members in the business organization publicly demonstrate their commitment to ITSM. They discuss the business problems to be solved by it. They show up, even if only briefly, at key staff meetings to speak about it. They may have their picture in a newsletter or e-mail that touts progress or status about the ITSM effort. They participate in steering committee meetings that guide the overall effort.

Expect that Senior Management may feel a bit awkward at this from the start. They most likely may not fully understand how ITSM will change the delivery model for IT. They may not be sure what events they should appear at or what they should say. For this reason, it will be important, through those very important organizational change efforts, to provide coaching and support for Senior Management stakeholders.

#9—Place Critical Importance on Metrics

Successful IT Service Management is highly process driven. Therefore, an ITSM solution involves the implementation of process and cultural changes to improve services and achieve business benefit. Most IT projects involve the implementation of hardware and software solutions. These can be seen and touched. The servers and software are either installed and running, or else it is assumed that something is wrong. Since the focus for ITSM is the implementation of processes and a new service culture, how does one know whether that implementation has been successful or not?

The answer is through metrics. A process cannot be considered implemented successfully unless proof is shown, through metrics, that it is operating correctly and meeting business need. Creating nice process binders and training staff in new processes is not a guarantee that efforts have been successful. While these are certainly tangible accomplishments, the real proof that improvements are happening can only be shown by regular and ongoing reporting of metrics that demonstrate what is truly going on.

For this reason, the following should be established as part of the ITSM transformation effort:

- Metrics should be defined for the overall effort to demonstrate that the goals of the ITSM transformation are being achieved

- Metrics should be defined for each process within ITSM to demonstrate the health and well-being of the process solutions

- A governance function should be included in the overall transformation project organization to collect and report on metrics on a periodic basis (usually monthly)

Identifying metrics is no small task. Some key guidelines for this might include:

- Aligning chosen metrics with the ITSM strategy and goals that will resolve the compelling business issues identified at the start of the effort

- Making sure that the metrics identified can be collected and reported on without excessive administrative overhead or cost

- Making sure that metrics are selected that directly indicate the business benefits to be achieved

- Choosing metrics that will allow management to act and make decisions versus those that simply report historical information on what occurred during the reporting period

- Documenting assumptions for how metrics are calculated and being used

- Creating an agreed reporting template or dashboard for presenting the metric results

Metrics fall into several types of categories that indicate different things. You will find that certain segments of management may be more interested in one type versus another. These are:

ITSM Transformation Metrics

These demonstrate whether the goals of the transformation effort have been achieved or not. They are typically of interest to project management and sponsors. An example might be something like "percentage increase/decrease in Severity 1 Incidents" if reducing major outages was one of the goals of undertaking the transformation effort.

Service Metrics

These demonstrate whether there are service issues occurring or about to occur. They are typically of interest to Service Managers, Service Representatives and Business Unit Representatives. An example might be something like a listing of planned service targets against actual results to indicate whether services are being provided in accordance with Service Level Agreements.

Process Metrics

These demonstrate whether ITSM process solutions are operating in a healthy state or not. They are typically of interest to Process Owners and Service Managers. An example might be something like "percentage of RFCs implemented without error" to demonstrate whether the Change Management process is functioning well.

Operational Metrics

These demonstrate activity and events within the IT infrastructure to gauge whether incidents are occurring or about to occur. They are typically of interest to operations and support personnel as well as others responsible for delivery of IT services. An example might be something like "CPU Utilization" for a server to determine if capacity and performance are at risk.

Metrics are absolutely critical to the ITSM effort and its ongoing operation. They are the only way to communicate true progress to Senior Management. They ensure that ITSM issues are recognized quickly and addressed in a timely manner. Without metrics, an ITSM solution is like a ship without a rudder. No one is sure where it is truly headed and whether it is really going anywhere.

If you are in a situation where Senior Management support is not in place, expect that support will take additional time to obtain. The following steps may be helpful in getting this support:

1. Establish a compelling business reason for why ITSM is needed

2. Establish a compelling reason for why ITSM is needed now (versus later)

3. Estimate key costs and timeframes

4. Estimate key benefits

5. Identify key stakeholders that can help push the message up through successive management chains

6. Establish a communications campaign to hone your messages and use your stakeholders to provide feedback (i.e. "Bill likes to hear it like this way . . .")

7. Schedule key meetings

8. Present your case

9. Close by getting not only agreement, but next steps to proceed

Other things to consider:

*   Assume that you have less than 4 minutes to present your case

*   Present no more than 3 main messages and make them simple and clear as complexity will result in confusion which will end up in rejection

*   Remember that agreement without next steps is equal to rejection—nothing will happen

*   Support your arguments with facts, key measurements or experience by other companies in similar industries

*   Do some research in advance—what does Senior Management see as key problems? What kinds of questions might they ask about your proposal? What presentation styles appear to work best with them?

*   Don't discount the power of tapping into Senior Management emotions and biases

Chapter

3

ITSM Transformation Overview

> "The secret of change is to focus all of your energy, not on
> fighting the old, but on building the new."
> —Socrates

Studies have shown that an IT Service Organization could achieve up to a 48 percent cost reduction by applying ITSM principles. (Microsoft ITForum Conference in Copenhagen, Denmark)

This chapter gets down to the nuts and bolts of how to move your organization forward towards implementing ITSM solutions and building an IT service culture aligned with where the business is headed. It focuses on the entire transformation effort from a 50,000 foot view. The goal is to orient you with the big picture first. The chapters that follow this one will then address each Work Stage that is presented here in much more detail.

The approach presented is based on practices and methods actually used at other companies. At its core, it follows two major lines of effort: building a solid ITSM foundation core within your organization while at the same time taking key actions to overcome service issues and deficiencies that will be noticed and recognized by the rest of the organization.

Guiding Principles have been used in pulling the transformation approach together. For the purpose of this chapter, the Guiding Principles presented here are very high level general statements about how an ITSM implementation effort should proceed. These have an impact on how the approach was organized and pulled together. The Guiding Principles and rationales used were as follows:

1. **The goal is to transform to an IT Service Management culture and operation, not implement individual processes or improve a process maturity score.** ITSM benefits lie in how the processes, tools and organization interlock and work together to meet business needs.

2. **The ITSM transformation effort should occur in waves with achieved benefits at the end of each wave.** Each wave should not last longer than 6-9 months. Projects

that run beyond that duration are harder to control in terms of scope, keeping personnel motivated and management interested. Corporate management has little patience for major efforts that span multiple years before any return on investment can be seen. If you cannot show results within the first wave, there is a major risk that the entire ITSM program will suffer a drop in funding and priority within the organization.

3. **The effort must be driven by a Program of Organizational Change.** ITSM represents a major mind set change towards an IT Service Culture. Processes do not happen by design and documentation alone. While IT is historically very good with implementations of technology, the real benefits will come from operating the processes with a set of service behaviours, not implementing new tools.

4. **The effort must balance strategic efforts with short term gains.** Short term gains are necessary to achieve some immediate wins, but you cannot lose sight of the longer term goals. If you do, you run the risk of developing point solutions that are disorganized and not ready for future business needs.

The above Principles have been highlighted here because they truly guide the strategies and solutions discussed later on. The entire transformation strategy and program organization have been built with them in mind. You will discover that the approaches recommended are targeted to provide initial wins without sacrificing long term strategy. They are targeted to maximize acceptance of the solutions being developed.

An Overview of the ITSM Transformation Approach

The overall approach uses a series of six high level Work Stages to accomplish the transformation tasks. This can be shown as follows:

ITSM Transformation Approach

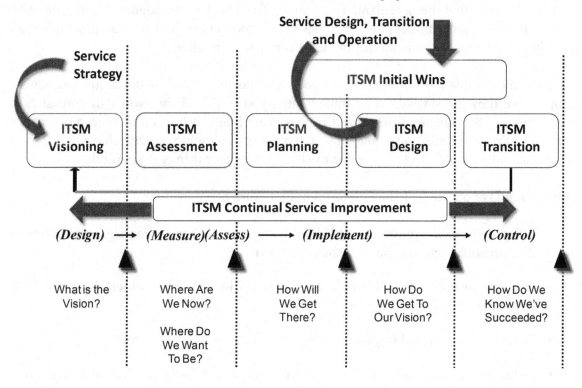

The above chart shows how the Work Stages flow to accomplish the transformation approach. The dotted line segments represent checkpoint periods within the transformation effort where key management decisions will occur to move forward to each successive Work Stage.

While not cast in stone, it is recommended that the overall transformation effort occur in repeatable waves. Each wave should not exceed 6-9 months in duration.

This entire cycle we will refer to as the ITSM Transformation Lifecycle. Do what you can within the first wave, then for each succeeding wave, circle back to the ITSM Visioning Work Stage and proceed to the right again. For initiatives that cannot be accomplished within a wave, think about breaking these down into smaller pieces, some which may be scheduled to be handled in succeeding waves.

Now let's take a high level look at what happens within each Work Stage.

ITSM Visioning Work Stage

In this Work Stage, the goal is to identify and agree the IT service vision, scope, and key business benefits that the organization is looking for. There is a tendency among many IT organizations to jump immediately into ITSM, hire consultants and start an assessment as quickly as possible. Try to avoid this mistake as much as possible.

The outcomes of this Work Stage provide powerful means to much better focus assessment efforts once they get started. More importantly, you will find that you will consistently return back to the outcomes developed in this Work Stage to build and keep momentum as things move forward. These outcomes will also serve to help you with those resistant to ITSM and those who claim they are just too busy with other things to spend time on it.

The key kinds of questions to be answered in this Work Stage include:

- What is the vision or future state for IT operations and service management practices within your organization over the next 1-3 years?

- What IT or business problems exist that might be solved by undertaking an ITSM implementation effort?

- Why should an ITSM implementation effort start now versus later?

- Who are the key sponsors, champions, movers and shakers that will carry the effort forward? Is there enough interest and momentum to make this happen?

- What benefits can be expected by implementing ITSM and when will these occur?

- What is the scope and impact of the ITSM effort? What things will change within the organization as a result of it?

- What kind of high level costs and labor might be involved if the organization moves forward with an ITSM implementation effort?

- Is there a relatively good understanding by key members of the organization about ITSM concepts, benefits and what it may take to implement them?

And finally the key question:

Does the organization see enough value in ITSM to move forward to the next Work Stage?

ITSM Assessment Work Stage

The mission of this work stage is threefold: identify what gaps exist between the current state organization practices and the planned service vision, what key actions need to take place to close those gaps and what Initial Win projects will be put into place for this first year. In Initial Win project is like a small project performed to accomplish a tactical task. An example of one might be implementing an IT Service Catalog.

The assessment will focus within four major areas:

Process—Reviewing the current state and maturity of processes within the organization and benchmarking them against ITSM best practices.

Technology—Reviewing the technologies currently in place and assessing how well they will support the future state service vision as identified in the previous ITSM Vision Work Stage.

Organization—Reviewing the current state organization and assessing how well it will support an IT service culture. This includes assessing the organization's readiness for change and willingness to undertake an ITSM implementation effort.

Governance—Reviewing current practices around IT service reporting, success factors, service and quality improvement initiatives, related metrics and data repositories to assess how well they will support an IT service culture.

Key questions to be addressed during the Assessment Work Stage include:

- How well does the organization perform compared to ITSM best practices?

- What actions should be taken to help the organization achieve its ITSM vision?

- When should these actions be taken? Why should we do any of these now?

- What might happen if we delay implementation of the recommended actions or simply do nothing at all?

- What Initial Win Projects should be undertaken this year that will provide noticeable and significant benefits?

- Will there be major impact to our existing IT organizations, technologies and control practices? If so, what kind of impacts might occur?

And finally the key question:

Now that we see what needs to be done, does the organization still wish to move forward to the next Work Stage?

As a last step, the assessment findings and recommendations will be reviewed with Senior Management executives. The outcome of this review should be an agreement on the recommended actions and a desire to fund and move forward in earnest with the ITSM transformation effort.

The assessment results will serve as a solid guide as to what needs to be done over time to move towards the desired ITSM vision. This is critical input towards the next Work Stage which is ITSM Planning.

ITSM Planning Work Stage

The main goal of the ITSM Planning Work Stage is to produce the detailed transformation plans and establish the overall program infrastructure in terms of personnel, organization and working standards. This is the stage where everything comes together as a working program prior to starting actual ITSM transformation work.

It goes without saying that a Planning Stage needs to come before transformation work gets started. Without this, the effort will wander aimlessly, confusion will reign and the objectives of the transformation program will most likely not be achieved.

Key questions to be addressed during the ITSM Planning Work Stage include:

- How will the program be organized to build a successful implementation program? How will this program be structured?

- What roles need to be in place to execute this program? What skills are required for those roles and which personnel will be assigned to them?

- How much labor will be needed to execute the program?

- Have we considered all costs for this effort?

- What key activities and tasks need to be accomplished to meet the goals of the implementation effort?

- What key Work Products need to be produced? When do these need to be delivered?

- How will project teams communicate with each other? How will program conflicts between teams be resolved?

- How will program status be reported on?

And finally the key question:

Are costs, labor and duration still acceptable, and if so, are we ready to start the implementation effort?

You may have noticed that this Work Stage is done in parallel with the Initial Win Work Stage. You will see later on that this Work Stage also starts with a planning step. That stage will consist of series of Initial Win projects. Therefore, there is good synergy to coordinate the planning pieces for each of them with the ITSM Planning efforts done in this Work Stage.

ITSM Design Work Stage

The goal of the ITSM Foundation Work Stage is to establish and agree the strategic goals, activities, policies, working procedures, responsibilities and tool direction for each ITSM solution and then design them. This is the stage where processes will get designed along with organization, technology and governance strategies. This is where the strategic ITSM Service Vision will be designed and fleshed out.

Similar to the ITSM Assessment Work Stage, the four areas of Process, Technology, Organization and Governance will be designed for each ITSM process. These form the four pillars of a holistic ITSM solution.

There is an important two-way relationship that exists between this Work Stage and the Initial Win Work Stage. The Initial Win Work Stage is highly tactical in that it is tightly focused on implementation of discrete projects. The ITSM Foundation Work Stage differs in that it has the long term view in mind and is very strategic.

The Foundation and Initial Win Work Stage carry a symbiotic relationship with each other. The Initial Win projects will have their solutions designed in accordance with the strategic decisions determined in the Foundation Work Stage. Likewise, the Foundation Work Stage will find that strategic decisions will be tuned and adjusted by the real world application of the solutions developed in the Initial Win Work Stage. Both these stages feed off each other quite well and it is why they are shown as executing in parallel.

Key questions to be addressed during the ITSM Foundation Work Stage include:

- What is the mission and guiding principles for each process?

- What critical performance indicators and key success factors should exist for each process?

- What process key activities, workflows, working procedures and policies should be in place for each process?

- What inputs and outputs flow from each process? How will processes work together to form a seamless service support and delivery infrastructure?

- What technologies will be used to support each process? How will these work together?

- What organizational roles, skills and reporting structures should be in place to support the processes?

- How will processes be rolled out to the organization? How will these be communicated to those impacted by them?

- What forms, templates and reports should be in place for each process?

And finally the key question:

Is there agreement on the design for the future state ITSM Vision and is the organization ready to implement this design?

ITSM Initial Wins Work Stage

Initial Wins are discreet projects that provide immediate benefits. These are tactical efforts that have specific goals that can be seen, touched and felt by those in the organization. Wins may involve any of the areas of Process, Organization, Technology or Governance. They can be done in a 2-4 month timeframe and they have a clear sense of accomplishment when completed.

This Work Stage runs in parallel with ITSM Planning, Foundation and Control Work Stages to make sure that Initial Win solutions are integrated with each other and aligned to strategic ITSM goals and direction.

Examples of Initial Win Projects might include:

- Implementing an IT Service Catalog

- Implementing Problem Classification Categories

- Designing a Configuration Management Database logical data schema

- Implementing an IT Chargeback Solution

- Establishing a standard risk analysis methodology for services with availability issues

As can be seen, the activities within the Initial Wins Work Stages are fairly straightforward. They are essentially a series of short term implementation projects that are taking place within each ITSM Process. At the end of each one, however, your organization will have gained some recognized benefits and taken a noticeable step towards the future state ITSM vision.

Many more examples of Initial Wins are shown in the Initial Wins Chapter that describes that Work Stage.

ITSM Transition Work Stage

The main goal of the ITSM Transition Work Stage is to implement the processes designed in the ITSM Design Work Stage and begin the regular cycle of reporting on their health and state. This is the Work Stage where the pedal hits the metal in terms of moving designed processes outside the implementation teams to an active state and implementing the necessary cultural and organizational changes necessary to support desired ITSM activities.

Example activities that can occur in this work stage include:

- Executing transition activities to new ITSM solutions

- Training personnel on how to use new processes

- Implementing and reporting on the process metrics

- Implementing new job descriptions

- Implementing organizational changes

- Recruiting and assigning personnel to ITSM job roles

- Activating ITSM Governance activities such as service reviews

At the end of this effort, all Initial Win projects will have been completed and the IT organization will be operating with ITSM practices and process controls. The ongoing cycle of metrics reporting, service reviews and service improvement initiatives will have been started.

Work Breakdown Concepts

The ITSM transformation approach uses a simple Work Breakdown model for organizing work activities and tasks. This model consists of Work Stages, Work Tracks, Work Tasks, Work Steps, and Work Products. These come together as shown in the following examples:

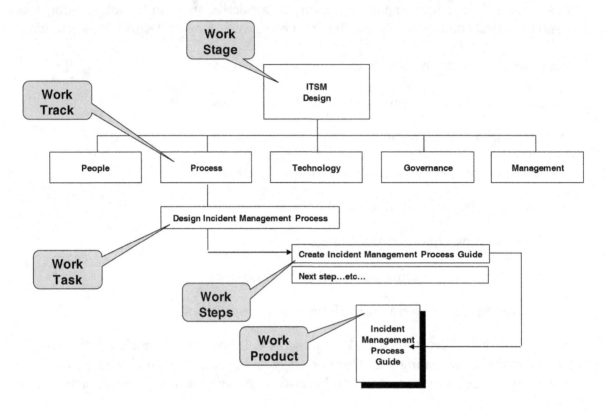

Work Stages

These represent the highest level grouping of implementation activities. They can be viewed as major phases in the ITSM transformation lifecycle. The six Work Stages are ITSM Visioning, Assessment, Planning, Design, Initial Wins and Transition as described earlier.

Work Tracks

All activities within a Work Stage are carved up into five Work Tracks. For each Work Stage, these will always be Process, Technology, Organization, Governance and Management. This ensures that each ITSM solution being put into place considers each solution holistically.

Work Tasks

These represent a key major implementation activity within a Work Stage. It helps organize the many implementation tasks in a logical manner. Work Tasks may start and complete at different times.

Work Steps

Each Work Task is broken down into smaller Work Steps. Work Steps proceed in a sequential order to accomplish a Work Task. A Work Step is also owned by one Work Track. Other Work Tracks may assist with it, but only a single Work Track will own it.

Work Products

These are ITSM implementation artifacts produced by Work Steps. An example might be a Process Guide or Service Report Template.

Ownership of each Work Task will be assigned to one of the five Work Tracks. Other Work Tracks may also assist the owning Work Track in completing assigned Work Tasks. To describe how the Work Tracks will assist with each Work Task, a RACI approach has been taken. For each Work Task a Work Track will be assigned ownership as follows:

R
Responsible, which means the Work Track owns the Work Task and has ultimate responsibility for its completion

A
Assists, which means the Work Track will directly assist the owning Work Track in completing the Work Task

C
Consults, which means the Work Track will provide advice, review and counsel to assist the owning Work Track in completing the Work Task

I
Informed, which means the Work Track is informed by the other Work Tracks about the Work Task and what was produced.

Detailed work break-down structures have been provided for each Work Stage in the ITSM Transformation Lifecycle. These can be found in the chapters devoted to each Work Stage. In addition, you can download these plans directly from ITSMLib from the ITSM Implementation library as tooling aids. Each plan describes a recommended set of roles and activities as well as an implementation organization structure for performing Work Tasks and Work Steps.

Process Concepts

For the purposes of this book, we will use the following definition for a process:

> A process is a connected series of actions and activities performed by agents with the intent of satisfying a purpose or achieving a goal.

Process names start with nouns. For example, a process name would be Incident Management or Capacity Management. Processes contain a number of elements that make them complete. These elements can are shown in the following model:

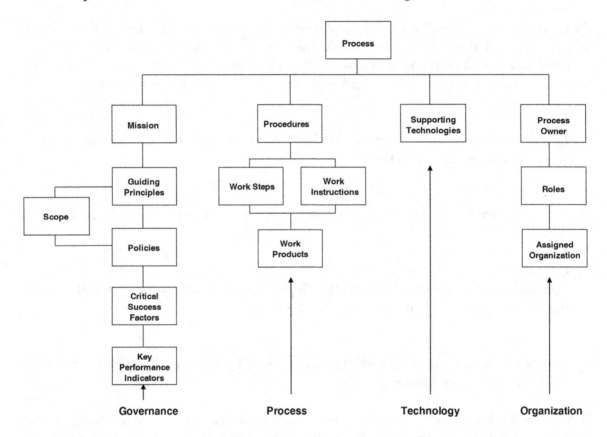

The model identifies all recommended elements that will make your processes effective and successful. Notice that these have been divided into four areas that match our Work Tracks: Governance, Process, Technology and Organization. Time has been taken here to identify these elements because these are what will be developed and built for each ITSM process as part of the transformation effort.

The bulk of the work for defining these will occur in the ITSM Design Work Stage. It will be the ITSM Initial Wins and Transition Work Stages where they get implemented.

A brief description of each of these elements is shown starting on the following page.

Procedure

This is an established method for carrying out a process activity. Procedures start with verbs. For example: *Detect Incident* or *Classify Problem*.

Work Step

This is a discrete grouping of Work Instructions that occurs within a procedure. Procedures may have one or more Work Tasks associated with them. Work Steps serve to break long strings of Work Instructions into easier to understand groupings. Work Steps also start with verbs. For example: *Enter Incident*.

Work Instructions

This is a detailed breakdown of sequential steps needed to execute a procedure or a Work Step. These also should start with verbs and should be sequentially numbered. For example:

1) Receive call or event
2) Triage the call/event
3) Validate that it is an incident
4) Press F4 on keyboard to open an Incident
5) Type in Incident description
6) Etc . . .

Work Products

These represent artifacts produced by Work Steps or sets of Work Instructions. Examples might be artifacts such as a Service Report or Process Guide.

Mission

This describes the main objective for a given process. It should typically be no longer than one sentence that succinctly describes why the process exists.

Guiding Principles

These consist of one or more statements of direction that will have major impact on how the process should be designed and/or operated. They are derived from basic beliefs, underlying culture or experience from within your company. They consist of a Principle Statement, Rationale and Implication description. Examples of these are included in the Chapter on ITSM Design Principles in this book.

Scope

This describes who or what will be impacted by the process. It may include departments within your company, certain customer segments, and users of certain kinds of business or IT services or a geographic region/business unit.

Policies

These consist of one or more guidelines, rules or principles that govern how a process will operate. An example might by a Release Policy that governs how releases will be built, named, packaged and distributed.

Critical Success Factors (CSFs)

These are one or more statements that describe what must happen for a process to be considered successful. An example might be: *Maintain Customer Satisfaction at High Levels.*

Key Performance Indicators (KPIs)

These are one or more metrics that will be used to measure whether Critical Success Factors have been achieved. They should be described in terms of a metric, calculation assumptions and a target range that indicates success. As an example:

Customer Survey Scores as calculated from the Monthly Customer Survey Database must fall between 8.5 and 10.0.

Supporting Technologies

These are tools and architectures that support the process. An example might be an Incident Management tool used to log and track Incidents.

Process Owner

This is an organizational role with overall responsibility for process deployment and operations. A more detailed description of this role is in the Chapter which discusses ITSM Transformation Organization.

Roles

These represent ownership groupings of tasks and activities that support a process. Examples of these might be a *Process Owner, Configuration Librarian* or *Service Manager.* A comprehensive listing of roles and their descriptions is provided in the Chapter on ITSM Organizational Roles.

Assigned Organization

This represents organizational entities within your company that have been assigned ITSM roles. An example might be a manager in the IT department who will be assigned the *Service Manager* role. This represents real world assignment of ITSM roles and responsibilities.

Chapter 4

Adapting to an ITSM Culture

"If you do what you've always done, you'll get what you've always gotten"
—Anthony Robbins—Advisor To World Leaders

You cannot just simply design and build your processes and ITSM solutions and toss them over the wall to the organization. This is a recipe for disaster. There is a very high risk that you will incur great resistance within the organization. The cause of rejection can vary from pure emotional reasons to political issues or simply frustration at having to learn how to do something differently.

Therefore, adapting to an ITSM Culture will require a program of Organizational Change. This is about getting to the hearts and minds of each company employee to embrace changes in culture and attitude towards an IT service culture. Perhaps you are asking yourself about the need for this. After all, IT works hard to get things done. They are always thinking of being helpful towards the business organization.

Take a look at your IT organization chart. Is it truly organized for service provision or more like a configuration database schema (i.e. Midrange organization, PC/LAN, Mainframe, Network, etc.)? If your Email service fails does anyone have day-to-day service responsibility end-to-end from servers to networks and down to employee workstations? If someone actually has that responsibility, are they truly given authority to reprioritize IT work on behalf of service?

As you look at the managers and key decision makers in your organization you may notice that they are grouped and aligned by technology silo. There might be the Director of mid-range operations for example, the Manager of desktop services, and the Vice President of network services and so on. In other words, all *key decision making* is done with no eye on end-to-end service, only technologies.

The largest ITSM hurdle to overcome is adapting the organization such that:

Key decision making is now done by services delivered to the business.

Many companies are unprepared for this and do not realize that this is where the true benefits will spring from. They are usually much unprepared for what this truly means to their organization. Almost any successful ITSM transformation effort started at any level of the company usually ends up in some Senior Executive's office sooner or later.

Many of those in IT will tend to treat this effort like a hardware or software solution. Simply hand it down to the IT staff and things will improve. When confronted with the need to address organizational change on an ITSM implementation effort, most IT staff will shrug it off feeling it is not necessary. They want to jump into design and solution work as fast as possible. On the management side, the attitude is simply "they will get it done—or there will be consequences" as a shortcut for avoiding this work.

Both attitudes will cause problems. If ignored, ITSM solutions stand a major risk that they will become nothing but nice looking binders on someone's shelf. Forcing ITSM on staff may get things moving for a short while, but this will serve to greatly lower morale. Even worse, the attention to this will quickly die out once there is a change in management or other priorities come into play.

Planning and executing an Organizational Change effort is not a simple task. In fact, most companies that have undergone an ITSM transformation will cite that they felt they spent 70-80% of the project working on these kinds of tasks! This is why a chapter has been dedicated for this. Realize that an ITSM effort is one of Organizational Change first and foremost. Building the processes and ITSM solutions is secondary to this. The best built ITSM solutions will fail if the organizational culture and habits do not go with them.

In looking at why so much time and effort is needed for Organizational Change, most companies will cite the reasons as:

- Educating staff and management on new practices and behaviors

- Holding frequent meetings to get stakeholders on board with ITSM solutions

- Gathering stakeholder concerns and requirements

- Crafting messages to teams, stakeholders, management and customers involved with or impacted by the implementation effort

- Dealing with resistance

- Communicating to people about ITSM activities and plans with the right messages at the right time

To make an ITSM change take root in the organization, at least four things must be in place:

1. A solid vision that addresses real business issues must be in place that is easy to grasp and understand

2. There must be a compelling reason to address ITSM now versus later

3. The path to get to the vision must be communicated in a manner where others can see how this might work within the business organization

4. Enough key decision makers, management, IT staff and business unit personnel agree with the vision, solution approach and path to get there.

All four items must be in place or the ITSM Program will fail. Here are some examples of what might happen when only one of the above is missing:

Key management personnel totally agree that ITSM is the right thing to do. They nod their heads. They might even congratulate you for your foresight and ideas. However there is no compelling reason to do this right now. You end up with a great meeting, but somehow the effort is never followed up on with staffing commitment or funding resources. *(Missing a compelling reason to do this now)*

Management and IT staff really like ITSM concepts and practices, but they're not sure how to make them happen. The organization is too entrenched in the existing ways of doing things. Therefore, the general feeling is that ITSM is a nice concept, but too high level and will never work. *(Others can't see a path for how this will get done).*

Management and IT staff have heard about ITSM but are not sure what benefits exist specific to the company. When asked about these, the IT department responds with a 20 page presentation on ITSM benefits and why this is important. There is agreement to start the effort. Funding may even be committed. However, the rest of the organization is not sure what they will really get out of ITSM without long winded explanations. This results in lack of participation and commitment on behalf of other business and IT units. *(No vision exists that can easily be communicated and understood)*

The ITSM Implementation approach presented in this book has been designed with Organizational Change in mind from the outset. Some of the ways you will see how this has been built in include:

Using Core, Extended, Advisor and Subject Matter Expert team Participation Roles to efficiently leverage a wide swath of the organization in agreeing and assisting with ITSM solutions. No one resists a solution that they had a hand in building, and this successful approach has worked many times in the past with other companies. (See the chapter on Transition Organization for details).

Using an Organization Work Track in each Work Stage of the ITSM Implementation Lifecycle to make sure there is a dedicated focus on Organizational Change throughout the transformation effort.

Developing an ITSM Communication Plan at the outset to address resistance and plan for effective communications among IT and business staff.

Inclusion of Organizational Change tasks that are built into each Work Stage, especially the first ITSM Visioning Work Stage, to set the vision, compelling reasons and initial agreement to move forward with an ITSM solution.

Developing an Organizational Change Strategy

In the Visioning Work Stage much preliminary work will be done to identify vision benefits and stakeholders. The ITSM Vision Communications Strategy can be constructed from this work. The purpose of this strategy is to identify:

- The key Vision messages

- The messages that each Stakeholder will receive

- The delivery channels to be used to deliver each message

- The timing for when messages are communicated

- The criteria for determining that the Vision has been accepted

Some considerations when working with the above are:

Vision Messages

Review all the benefits of the Vision and hone these into sound bites that sound meaningful and attractive to your organization. Keep things to brief and simple statements. If it can't be communicated in less than 30 seconds, it is probably too wordy or complex and will not be easily understood by others.

Stakeholder Specific Messages

You may need to create slight variations of your sound bites to better address individual concerns and issues. Use the survey output that will be done in the ITSM Vision Work Stage as input for this activity. Map each message with each Stakeholder. It is helpful to inventory your sound bite messages where you can easily find them to insert into presentations and other communications.

Delivery Channels

Determine how each message will be delivered and over what media. As an example, you might plan to deliver a series of planned presentations. These will be delivered via presentation slides. You may have other events such as a brown bag seminar for employees or a monthly newsletter. There is no one correct answer for these. Examine what is available in your organization and leverage it to get your points across.

Message Timing

Plan specific time frames for when key messages are delivered. For example, you may decide that the first wave of communications will occur in 2 weeks and only serve to create awareness about the overall program. The following week may have a second wave of communications that provides more specifics about the program tailored to different stakeholder groups.

Lay out a schedule and plan for which messages will be delivered and when. Remember that timing is everything. It is just as possible to over communicate rather than under communicate. Find the right balance that informs others about what is happening at the right time to build and keep the momentum up.

Acceptance Criteria

Determine the conditions by which you get the feeling that the Vision is accepted by each stakeholder. Although it is impossible to read minds to confirm this, there may be other means by which you can gauge acceptance. Some of these include:

- Stakeholders demonstrate willingness to join and help out in the effort (i.e. "Count me in . . .").

- Stakeholders demonstrate willingness to commit funds and resources.

- Stakeholders demonstrate commitment to promote the program.

Also make sure your meetings with stakeholders are two-way. That is, don't just communicate your vision, but ask for lots of feedback. What do they think about the vision? What challenges do they see with it? Would they address things differently? The kinds of answers you get to these questions will indicate the level of buy-in they might have.

Be sure you test the level of acceptance as much as possible. Some may nod their head and look like they are in agreement. They may just be acting polite but are not really bought into your Vision. In some cultures, such as in Japan, it is very typical for most people to nod positively as you discuss the Vision. You may interpret this as acceptance when within their culture it really means "I have heard what you have said . . ." which is not the same as "I agree with your Vision . . ."

Identifying Vision Impacts

An early step in developing the Vision Communications Strategy is to review the Vision that is being developed and identify the impacts of that vision in terms of who will be impacted by the Vision itself. As the Vision and the Vision Approach become developed, you need to review what is being constructed. Break down the vision into its core elements and determine what kinds of impacts might exist.

The following table provides an example of how you might review your vision for impacts on your organization.

ITSM Vision Organizational Impact Table Example

Impact Area	Potential Impacts
Customers	• Added value that will occur • Changes in customer satisfaction • Adding new customers • Reporting service to customers
Organization	• Must organize to deliver good service • Decision making prioritized by business and service needs
Employees	• Changes in how people are rewarded • Formalizing job roles and skills • Customer driven mindset • Need to formalize service communications • Training and skill sets
Process	• Changes in how role of processes are viewed and operated • Changes in how processes are monitored and reported on • Consolidation of similar processes • Single process focus across stovepipe organizational pillars
Technology	• Changes in how tools are selected • Changes in how tools are integrated • Architectural Changes
Governance	• Collection and reporting on services • Establishment of service targets • Reporting on service quality
Culture	• Service and Customer Orientation • Understanding the Customer • Prioritizing based on Customer Needs

Note that in the above table, the impacts described are very broad and general. In the ITSM Visioning Work Stage, you will be looking for the types of impacts that will need to be considered. You are not designing and answering how those impacts will be addressed. The goal is to understand the potential scope and breadth of what implementation of the Vision will mean to the organization.

Initial Stakeholder Identification

Another step in developing your communications strategy is to begin reviewing your organization and examining which business units and key personnel might be affected by the potential impacts that have been identified. These will become the Stakeholders for the ITSM Transformation Program. The Stakeholders may fall into one of a number of Impact Categories:

Customers

Customer and customer groups that will receive ITSM services under the Vision that is being developed and who must be made aware of how services will be organized, managed and delivered.

Senior Management

This includes key Senior Management personnel that have to make organization decisions and deal with transition and process changes to meet ITSM objectives. They need to be aware of, and possibly directly involved in the direction of the ITSM Vision.

Middle Management

This includes management personnel responsible for business units and departments that will be delivering or receiving IT services. They need to be aware of, and possibly directly involved in the direction of the ITSM Vision.

IT Delivery and Support Staff

Awareness must be created among IT Staff responsible for service delivery and support to gain their early buy-in to the Vision.

IT Technical Architecture Staff

Those involved with technical architecture decisions and/or major technical purchases should also be aware of the Vision and how this may impact their decisions and activities.

Sponsors

Key Senior Management personnel that may have to supply budget and funds for the ITSM transformation effort who will want to understand and agree the Vision and the Vision approach.

Users

Business staff that uses the IT services being provided who need to be aware of pending changes in how services will be organized, managed and delivered.

Vendor Suppliers

Outside 3rd party vendors that currently supply services to IT that should be aware of pending changes in how services will be organized, managed and delivered.

IT Development Staff

Awareness must be created with IT Development staff that might see impacts in how IT solutions are turned over and integrated with deployment and service operations.

Change Agents

This includes the ITSM Vision Team along with other employees that may soon be involved in building the ITSM solutions in the later Work Stages.

Champions

This represents individuals with strong opinions about how IT services should be managed and delivered and who are willing to participate in forming and pushing the Vision out to others in the organization. As an example, there may be someone from the User community who is willing to step up and spend time to promote and push ITSM solutions within their own organization.

Stakeholder Initiatives

This relates to other similar efforts that may already be occurring in the organization that might overlap an ITSM solution. An example might be another Change Management initiative that is occurring in another business unit or a corporate quality initiative program such as Hoshin, TQM, CMM or Six-Sigma.

It is very helpful to seek these out to identify where these efforts are, what their objectives are and how there might be impact with the Vision being developed. They can become powerful allies and change agents on your behalf.

Assigning Stakeholders to Team Participation Roles

Using the Impact Categories as a reference, you can begin to dig out all the organization charts and related documentation that might exist within your company. Start to pinpoint which organizations and personnel may be impacted by the Vision as it gets developed. Begin to think about how those individuals might participate in the Implementation Program as it moves forward.

Each Stakeholder should be assigned a Participation Role. Participation Roles identify how each Stakeholder will interact with the ITSM Transformation Program. Some Stakeholders may directly work full time with the program. Others may only provide expertise or be called on to provide key decisions. The recommended Participation Roles are:

Core Team

This represents people who will work with the program on a full time basis to build the ITSM Solutions.

Extended Team

This represents people who will participate part time in representing their business unit's stake in the ITSM solutions being developed to obtain agreement on them. This group will become the users of ITSM solutions.

Advisor Team

This represents those who will need to be consulted for key decisions and agree ITSM solutions. These people may also be sought out to identify others that should be involved in the Transition Program.

Subject Matter Expert

This represents those people with special expertise who will assist the Program as needed. This might include expertise in a technology area, business unit function, ITSM processes, organizational change or any other subject area where assistance is needed.

Expect that Stakeholder involvement will be fairly fluid. It is not atypical that Stakeholders may come and go throughout the life of the transformation effort. Therefore, management of Stakeholders is an ongoing activity throughout the ITSM Transformation Lifecycle.

More complete descriptions of Participation Roles are provided in the chapter on Transformation Organization.

Summarizing Vision Impacts and Stakeholder Information

Much of the information pulled together on Vision Impacts and Stakeholders can be placed in a summary table for easier reference. This table might look like the following:

Organization	Role	Impacts	Stakeholders	Participation
Sales Department	Customer	• Reporting of services • How complaints will be escalated	• John Smith (Sr. VP Sales) • Bill Hand (Sales Ops Mgr)	• Steering Group • Advisor Team
Enterprise Architecture	IT Technical Architecture	• Tool selection • Tool Integration	• Jim Adams (VP of IT Architecture) • Narendra Ushma (IT Architect)	• Advisor Team • Extended Team

The table above shows an effective way to tie your organization entities, impacts and key stakeholders together quickly at a high level. This table will be useful as a starting point for the Stakeholder Analysis activity.

In the table above, note that one person has been considered for the Steering Group. This is an oversight entity to identify Senior Management personnel who will oversee the entire ITSM Program.

Conducting Stakeholder Analysis

Stakeholders are those who will have a stake in the ITSM Transformation program and its results. They also include those who need to change any behaviors and actions as a result of the ITSM solution being developed. It is absolutely critical to understand and manage your stakeholders at all times throughout the lifecycle of the transformation effort. Failure to do can easily result in resistance to implementing changes, program delays or even serve to kill the entire effort.

A Stakeholder Map will be developed that monitors and manages your stakeholders throughout the life of the transformation effort. In the ITSM Visioning Work Stage, you will be building the initial map that will be enhanced in the later work stages.

The recommended steps to accomplish this are:

1. List your key stakeholders
2. List their title and organization they work in
3. Identify the Impact Category they belong to
4. Identify the priority they hold towards accomplishing the Vision Objectives
5. Identify their current stance on the ITSM Vision
6. Identify their influence role within their organization
7. Identify WIIFM (What's In It for me?) for each stakeholder
8. Identify their candidate role within the Vision Team
9. Identify issues and concerns about each stakeholder
10. Identify what you would need from each stakeholder
11. Identify what specific actions are needed with each stakeholder

Let's take a look at a few of these steps in more detail.

Identifying Stakeholder Priorities (Step 4)

Here you will be identifying the priority of the stakeholder in terms of relevance to accomplishing the Vision objectives. This is to make sure that the team is spending time appropriately with each stakeholder. For example, you would not want to prioritize time and effort on working with a stakeholder who might have little influence or decision making authority.

The following table identifies some recommended priorities and their criteria:

Priority	Criteria
Immediate	Someone you think is essential to contact immediately because of the critical nature of their support or influence on others. These are the first people you want to contact about your change.
Soon	Someone whom you need to contact soon, but maybe need the help of another stakeholder in order to influence them. Also, you may need to gather additional information before contacting them.
Later	Someone who for any number of reasons does not need to be immediately involved and perhaps it's better if they are not immediately contacted about the change so that you know how to involve them at the right time, in the right way and for the right reasons.
Unknown	Someone for whom you are not quite sure fits into any of the above priorities.

Identifying Stakeholder Current Stance (Step 5)

This identifies your best opinion on how supportive you believe the stakeholder to be towards the ITSM Vision and overall effort. Are they supportive? Are the resisting? The following table identifies some recommended stances and their criteria:

Stance	Criteria
Negative	Will not change their own behavior and they will publicly and actively work against the ITSM effort.
Against	Does not believe in the ITSM Program and will not participate in it nor make any changes as a result of it.
Neutral	Is against the ITSM Program but will hold off participation and changes for it until it appears the general organization will move forward with it.
Positive	Willing to participate in the ITSM Program and make the necessary changes to meet program objectives.
Champion	Willing to actively promote the ITSM Program and lead others to make the needed changes to meet program objectives.
Unknown	Stance is unknown in relation to the ITSM program.

Identifying Influence Roles (Step 6)

This identifies your best opinion as to how influential the stakeholder is towards the ITSM Program. For example, the stakeholder may be a key decision maker that can get things done. In another example, they may act as an influence to convince others. It is important to accurately recognize where each stakeholder is with this.

Do not let organization titles and hierarchy levels blind you to the real influence role that a stakeholder may have. For example, that overbearing Vice President in one of your business units may actually be making all their decisions based on input from the meek and quiet Assistant Manager that works for them.

The following table identifies some recommended influence roles and their criteria:

Influence Role	Criteria
Decision Maker	Will have significant say over direction and content of the ITSM Vision. Is able to back up decisions with funding and resource commitments.
Influencer	Influences others who are authorized to make key decisions. Will have significant say over direction and content of the ITSM Vision because of influence versus having funding and resource commitment authorization.
Controller	No significant influence or decision making authority, but manages and leads others once convinced about the merits of the ITSM Vision.
Information Provider	Has no influence or decision making authority, but is adept at keeping their ear to the ground and reading the tea leaves (as it were) about how the organization might respond and who else should be consulted. May very adept at politics within the corporation.
Subject Expert	Has critical expertise in a subject area that is impacted by or contributes to the Vision and its approach. Usually consulted for input. Stakeholders in this category may need to be consulted by others on the Vision Team to gain insight on an issue or subject matter.
Participant	Little or no influence, but will be a participant in the ITSM Vision once directed to do so.
Unknown	Influence role is unknown.

Identifying WIIFM (Step 7)

Identification of the WIIFM (What's in it for me?) for each stakeholder is key towards gaining their buy-in and acceptance. For a variety of reasons, each stakeholder has their own desires, wants and needs that they operate with. It is important to understand what these are and how they may relate to the ITSM Vision and Transformation Program overall.

Addressing these accurately with each Stakeholder will result quite favorably when it comes to overall acceptance of the ITSM Vision and willingness to accept the results of the Transformation Program. The more that everyone can see a stake in the Vision, the more successful you will be at leading your organization towards it. Likewise, those that do not see a stake in the vision will soon move onto other priorities and may even work against you if they feel the Vision runs counter to their own objectives and needs.

Determining the WIIFM for each stakeholder will be done mostly by meeting with them individually and flushing out their concerns and interests. By identifying the Impact Category for each stakeholder which you did in an earlier step, you can get a ready start towards what some of these might be.

The table below provides some guidance towards this. It lists each of the Impact Categories discussed earlier and identifies possible wants and needs that typically are found in that Impact Category:

Impact Category	Typical Wants and Needs
Customers	• Value for money spent if charged for IT services • Better service—can do work faster and more efficiently • More attention from IT • More influence over IT plans and activities
Senior Management	• Meeting corporate quality and improvement initiatives • Improving service • Getting ready for a major business initiative or change • Preventing calls from other business units about IT service issues • Demonstrating compliance with regulatory requirements • Understanding of costs for providing IT services
Middle Management	• Meeting business unit targets and objectives • Keeping staff and direct reports motivated • Preparing for a major business initiative or change • Preventing calls from upper management about IT service issues • Keeping IT costs in line with budget targets and objectives
IT Delivery and Support Staff	• Recognition for work performed • Eliminate firefighting • Working smarter instead of harder and longer • Reducing time spent on non-value work

Impact Category	Typical Wants and Needs
IT Technical Architecture Staff	• Major support for IT Architecture standards and initiatives • Major support for IT governance • Recognition for technical solutions that support desired processes and business objectives • Increased visibility for architecture role and activities
Sponsors	• Return on investments made in ITSM • Successful implementation program • Meeting program objectives
Users	• Get work done without service disruptions • Get work done without spending time dealing with IT technical issues and problems • Better communications about service outages and root causes for them
Vendor Suppliers	• Better relationship with your business organization • Opportunity to increase services or length of service contracts • Clear definition of service roles and responsibilities • Clear roles and responsibilities for handling problems and incidents to reduce finger pointing and blame
IT Development Staff	• More efficient process for placing IT solutions into production • Increased business satisfaction with new IT solutions • Clear roles and responsibilities for handling problems and incidents to reduce finger pointing and blame between development and operations • Ensure deployment and operational considerations baked in as solutions are developed versus scrambling at the end before deployment
Change Agents	• Recognition • Chance to learn something new • Work on interesting projects • Make a positive difference in the organization
Stakeholder Initiatives	• Gain increased support and sponsorship for initiatives already underway • Share like ideas and improve solutions already under design • Increased influence and visibility for the initiative being worked on
Champions	• See desires turn into actions that make a positive difference • Gain visibility and recognition for opinions and foresight • Desires to see things done in the best possible way

An important source not to be overlooked is the Customer Needs Survey, ITSM Focus Survey and the ITSM Benefits Survey which will be done in the ITSM Visioning Work Stage. A review of the survey results can help you determine where people put a priority on what they would like to see and get out of the Vision and the Transformation

Program. Items from these surveys that were placed at higher priorities can be added to the lists above.

Copies of these surveys can be found on ITSMLib in the ITSM Implementation Library as tooling aids.

The advantage of having the above list is that you now have points for discussion as you meet with each stakeholder. Once you have identified the WIIFM that you feel is appropriate, you can then update the Stakeholder map.

The example table on the following page illustrates how all of this can come together into a Stakeholder Map.

Stakeholder Map Example

Name	Title	Category	Priority	Stance	Influence	Role	WIIFM	Concerns	We Need	Actions Needed
Jim Adams	VP of IT Architecture	Senior Management	Immediate	Neutral	Decision Maker	Advisor	Reinforces architecture Visibility and control	Complexity of effort Might result in major tool changes	Act as champion and provide Extended Team members for the effort	(Mm/dd/yy) Meet to discuss participation
Bill Hand	Manager, Sales Operations	User	Immediate	Positive	Influencer	Extended	Better service for his users	Generally positive	Needs to convince John Smith to get on board	(Mm/dd/yy) Meet to discuss potential strategy for how might communicate to John
John Smith	VP, Sales	Customer	Soon	Negative	Decision Maker	Advisor	Unknown	Thinks this effort is a waste of time	Need to get him on board	(Mm/dd/yy) Schedule preliminary talk when Bill is ready
Mark Carter	VP, Service Delivery	Sponsor	Immediate	Positive	Decision Maker	Steering Group	Better service delivery Meets needs for customer focus	Coordination	Funding and walking the talk	(Mm/dd/yy) Conduct preliminary review of the vision and find out his interests and concerns

Name	Title	Category	Priority	Stance	Influence	Role	WIIFM	Concerns	We Need	Actions Needed
Mary Jones	Business Representative	Customer	Soon	Unknown	Participant	Extended	Unknown	No one on team knows her stance or needs	Need her support since she will represent 5 different business units	(Mm/dd/yy) Need to discuss participation
Jason Adams	VP, IT Operations	Senior Management	Immediate	Positive	Decision Maker	Advisor	Stop firefighting Increase service quality Work with best practices	Transitioning to ITSM without increasing staff head counts	Act as champion and provide Extended Team members for the effort	(Mm/dd/yy) Meet to review transition approach and identify additional extended team members from operations

Dealing with Resistance

Recognize that it is normal for people to hate change. We are all programmed by nature to act in this way. This goes back to our early cave-person days where we were programmed to act suspiciously of any change in weather, neighboring tribe, or hunting situation as a survival mechanism. In addition, those of us who are older in IT have spent a lifetime doing things a certain way with a reasonable amount of success in the past. Why bother to change now?

From this, you should recognize that people go through different stages in accepting change. They may be initially suspicious or hesitant to do things differently, but with the right exposure can adapt over time. The trick of your ITSM Communication Strategy is to provide the right exposure to overcome these kinds of barriers and gain acceptance.

Think about this scenario:

> The ITSM Transformation team goes offsite and develops ITSM processes and procedures. They then document these, return to the work site and get management to immediately make sure everyone follows what was documented.

What do you think the chance for success will be in the above scenario? Let's look at another:

> The ITSM Transformation Team drafts a Vision for how IT should operate. They then review this with key business and IT stakeholders. While solutions are being developed, a series of brown bag sessions and seminars are held over time. The first merely states why this is important. The next expand upon this and present a few additional concepts. The next involve a little game and role playing. Following this, people are trained in key concepts. Then an internal certification program in ITSM solutions is conducted. Over time, ITSM solutions are brought in piece meal and people are given a chance to provide feedback on their use and success.

In the first scenario, there is no preparation for change. Everyone is told to just "do it". There is a high risk that there will be much resentment among staff doing something "someone else's way".

In the second scenario there is at least a shot for success. People were introduced to new concepts over time and in small chunks. They had a chance to try out solutions ahead of time. They could even provide feedback and recommendations to what was put into place. They feel part ownership of the solutions and will therefore tend to not resist them.

This leads to several important tips when working on ITSM transformation efforts:

- People respond better if they had a hand in building/recommending the solution

- People respond better if introduced to new concepts in small chunks over time

- People respond better if allowed to experience new things in the form of role playing or game playing

Another consideration is to realize that people accept and process change in many different ways. I've somewhat likened these to the stages people go through when told extremely bad news (like you're about to die):

1. Denial ("Won't happen to me" or "Just another IT fad—it will soon pass")

2. Fear ("What does this mean to my job? Will I no longer be top dog around here?")

3. Pity ("Why are they doing this to me? Don't they know how hard my job already is?")

4. Bargaining ("Okay, but I'll only manage to two indicators and no OLA targets . . .")

5. Acceptance ("Great! I'm in!")

Another way to recognize and categorize resistance might be summarized by the following chart:

Resistance Type	Issue	What They Say	What They Do
Complacent	People do not understand that there is a need for ITSM	We are the best Nothing can stop us Don't fix what isn't broken We know what we are good at We've always been successful	Pat each other on the back Get less cost conscious Take their eye off the ball Promote past history of IT
Denial	People hope ITSM will go away or do not believe it is theirs to do	This is the fault of someone else We can't do that because . . . This happened before and we got by okay . . . Good point but you don't understand . . . . Our business is different We're too busy with other priorities	Finger-point and blame Become overly optimistic Reject key facts, metrics Pass the buck Look to past achievement
Confused	People understand ITSM but not how to implement it	How did we get in this mess? Where is this all going? Who, what, where, why . . . ? Too many initiatives around here Hire a consultant?	Look for direction Don't fully complete everything Adopt the latest fads Endless speculation and rumors Generate lots of overlapping or conflicting initiatives Constantly jumps into whatever sounds like the Holy Grail . . .

Learn to recognize these. People can accept change if given a chance over time. Realize that those who initially resist may become more accepting if given the right levels of exposure over time. They may follow the stages of acceptance as described earlier. Understanding where people are coming from can greatly help you lead and overcome resistance when it rears its head.

Defining an ITSM Communication Plan

The Communication Plan is the strategy document that will be used to direct and guide the ITSM Communications program. It will be constructed, managed and executed by the Organization Work Track. The plan itself defines the specific communications to be prepared and delivered, the timing for each communication, and the overall series of communications campaigns.

The plan will build upon the detail developed in the ITSM Vision Work Stage from survey results, the planned ITSM service vision and Stakeholder Analysis work. For each campaign, the plan identifies the communication media, timing, applicable messages for that media, and audiences to be targeted.

The following chart shows an example of first defining the campaigns and the proposed delivery channels to be used with them:

ITSM Communication Campaigns and Message Channels

Delivery Channel	ITSM Awareness and Understanding Campaign	ITSM Understanding and Positive Perceptions Campaign	ITSM Installation Program Campaign	ITSM Follow Up Campaign
E-Mail	X			
Voicemail	X			
Newsletter	X	X	X	X
Memo/Letter	X	X		
Presentations and staff meetings	X	X	X	
"Ask Your Question" voicemail box		X	X	
Management scripts	X	X	X	
Intranet	X			X
One-on-one conversations	X	X	X	X
"Day in the Life" video		X	X	
Education and training sessions			X	X
Mentoring and coaching			X	X

Performance expectation setting			X	X
Performance reviews				X
Rewards and recognition		X	X	X
Resistance management	X	X	X	X

For each campaign, objectives should be defined. Using the example campaigns above, some example key objectives might be as follows:

Campaign: ITSM Awareness and Understanding

- 100% of directly impacted personnel will be aware of the ITSM Transformation Program.

- The majority of company Senior Management will be able to explain the ITSM Service Vision and ITSM Program objectives in their own words.

Campaign: ITSM Understanding and Positive Perceptions

- Over 80% of the directly impacted personnel will be able to explain the ITSM Transformation Program and its key objectives.

- The majority of company Senior Management will view IT Service Management as necessary to the long-term competitive positioning and survival of the organization.

- 75% of impacted personnel will have voluntarily participated in at least one Brown Bag Lunch presentation about the ITSM Transformation Program.

Campaign: ITSM Installation Program

- 100% of directly impacted personnel know the required ITSM training schedule applicable to their position.

- 80% of all inbound communications are answered within 4 business hours.

Campaign: ITSM Follow-Up

- The majority of directly impacted personnel can describe at least one benefit or success story associated with the ITSM Transformation Program.

- 90% of all inbound communications are answered (or acknowledged, if the answer must be researched extensively) within 2 business hours.

Once goals are established, the key types of information needed by each audience type for each campaign can be planned. This can be drawn from the Stakeholder Analysis. An example is shown on the following page:

Example Campaign: ITSM Awareness and Understanding

Audience	Information Requirements	Specific Information Requirements
Senior Management	Overview of project Project goals and timing	Interrelationship and/or impact on other corporate initiatives and financial position Project priority Project leadership and staffing Project status and issue discussion
Middle Management	Overview of project Project goals and timing	Their personal and departmental responsibilities Impacts on their department and timing Impacts on their personnel and timing Who to contact for answers
IT Delivery and Support Staff	Overview of project Project goals and timing	Impacts to their position and timing Who to contact for answers
IT Technical Architecture Staff	Overview of project Project goals and timing	Positions to be impacted Impacts on the positions and timing Ways they will be informed and involved

After the delivery channel and messages have been identified, you can then lay these out over a planned timeframe. The following chart shows an example of this.

Media	By	Messages	Audiences
Memo to executives and directors	mm/yy	#1, #2, #3	Senior Management
Memo to all employees	mm/yy	#1, #2	All audiences except customers
E-mail (2)	mm/yy	#1, #2	All audiences except customers
Voicemail (2)	mm/yy	#1, #2	All audiences except customers
Newsletter (2)	mm/yy	#1, #2	All audiences except customers
Intranet	mm/yy	#1, #2	All audiences except customers
Management scripts	mm/yy	#1, #2	Senior and Middle Management
Staff meeting—executives	mm/yy	#1, #2, #3	Senior Management
Staff meetings—managers/supervisors	mm/yy	#1, #2	Middle Management
Staff meetings—employees	mm/yy	#1, #2	IT Service Delivery Providers

The above can then be fashioned into a work plan by the Organization Track. An example of a Table of Contents for an ITSM Communications Plan is shown below.

ITSM Communications Plan

1.0   Preface

2.0   Overview

3.0   ITSM Vision Impact Summary

4.0   Target Audiences

5.0   Stakeholder Analyses

6.0   Campaign Definitions

7.0   Campaign Objectives

8.0   Key Messages

9.0   Delivery Channels

10.0   Campaign Work Plans

Chapter
5

Transformation Organization

"We should work on our process, not the outcome of our processes."
—W. Edwards Deming

Before embarking on an ITSM transformation, care should be taken to organize appropriately for the efforts to follow. A recommended organizational solution is described in this chapter that will allow you to put a highly structured Program into place to guide the overall ITSM effort. This program maximizes the need for solution design, solution communications and solution acceptance. The model presented here has been used for very large global companies, however feel free to adapt for its concepts if your efforts will be smaller in scope.

Program Roles and Assignments

Every Program Team Member who works on this effort will be assigned the following:

- Team Participation Role

- Team Implementation Role

- Work Track Assignment

- One or more Initial Win Assignments

The above assignments are used as a basis how the implementation team organization structure has been put together. A brief description of these:

Participation Role

This indicates the level of assistance that the team member is providing. This chapter will review Participation Roles in more detail. These can be summarized as:

Core Team Member—Provides full time team participation dedicated to implementation work.

Extended Team Member—Represents the interests of a business or IT unit—could be full or part time.

Advisor Team Member—Provides key decisions, approvals or appoints Extended Team members.

Subject Matter Expert—Provides expertise in a specific area or technology as needed.

Implementation Role

This indicates the implementation role the team member will play during the transformation effort. Examples of an implementation role might be *Process Architect*, *Tool Developer* or *Trainer*. A listing of the Program Implementation roles and more detailed descriptions around these is provided later on.

Work Track Assignment

This indicates which Work Track the team member will be assigned to. The Work Track concept was discussed earlier in the ITSM Transformation Overview chapter. This will be a choice of *Process, Technology, Organization* or *Governance*.

Initial Win Assignment(s)

Many team members will also be assigned to one or more Initial Win efforts. The Initial Win concept was discussed earlier in the ITSM Transformation Overview chapter. Remember that Initial Wins represented smaller implementation projects that serve to bring high benefits within the first year at acceptable cost. As an example, a given team member may be assigned to something like *Implement Service Catalog* and *Implement OLAs* which were identified as Initial Win projects to be done this year.

Here are some example scenarios for how this might work together:

John is *Core Team* member who works as a *Process Architect* in the *Process* Work Track. He is also part of the project team working on the *Implement Service Catalog* Initial Win.

Shirley is an *Advisor Team* member who is an *Advisor Team Stakeholder*. She is in charge of all IT Technical Support. Shirley reviews and approves all key Work Products produced by those working in the Technology Track. Shirley is considered part of the *Governance* Track. (Note that many decision makers will fall into this Track). Shirley is not assigned to any specific Initial Win, but will review any Initial Win that involves a technology decision.

Anita serves as an *Administrative Analyst* in the Program Office. She works full time in this role. Therefore she is a *Core Team* member working in the *Governance* Work Track. She is not specific to any Initial Wins.

Bill is the CIO. He will make many of the ultimate decisions and funds the implementation effort. Bill also serves on the ITSM Steering Committee. Therefore, Bill participates as an *Advisor Team* member who serves as a *Steering Group Member*. By default, he is part of the *Governance* Work Track. Bill is not assigned to any Initial Win.

Sarah works in the Finance department. She sits in key meetings to make sure Finance technology needs are represented. Sarah is an *Extended Team Member* who is also part of the *Technology* Work Track. She is not assigned to any specific Initial Win effort.

You may be wondering why this is all necessary. The answer is that on large transformation teams, there is great benefit in establishing common roles and responsibilities. This is even more crucial if your transformation effort is global. Consider this item:

Bill works in the Corporate Affairs Personnel Division and has been assigned to the ITSM Transformation Program.

We now know where Bill works, but what is he doing on this project? Who is he working with? Where does he fit in? Even if Bill explains all this, what words will he use? Will they be the same if you ask his Manager? How would this work if you find another team member? Would they describe their participation totally differently? Consider this next item:

Bill is a Core Team Member working as an Organization Change Analyst on the Organization Work Track who is assisting with the Implement Personnel Service Incentives Initial Win effort.

In that one sentence, how much do you now know about Bill and what he is doing in the Transformation Program? How would the above understanding change if the Corporate Affairs Personnel Division unexpectedly undergoes an organization name change and becomes the Consumer Affairs Human Resources Division? (Hint: It doesn't!).

Having standard roles on longer efforts such as an IT Service Management Transformation goes a long way towards minimizing confusion over roles and responsibilities. It also prevents whatever organization chaos that might exist in your company from corrupting your efforts. In a global effort, it allows everyone involved to quickly understand each other even though cultural and language barriers might exist between company organization units.

Program Organization Structure

Having identified the overall organization roles and assignments, it is a short step to build the appropriate Program organization structure that will be used to transition to ITSM solutions. Such a structure is presented here. This structure has been used in the past for a large global ITSM transformation program. It has been chosen as an example since a large global effort presents many additional challenges and issues.

Keep in mind that this represents only one possible way to organize your ITSM efforts. You should feel free to adapt this model to whatever will fit your organization and size. The organization structure can be pictured as follows:

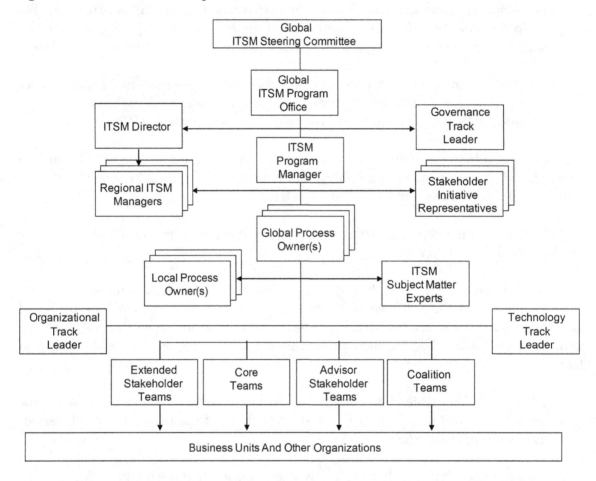

This structure uses many elements that are recommended to be in place as you execute your program. Some additional entities have also been added that are unique to a large global effort. Let's take a look at each organizational entity in more detail:

Global ITSM Steering Committee

This consists of Steering Group members who will guide and oversee the direction of the entire transformation effort.

Global ITSM Program Office

This entity consists of an ITSM Program Manager and Administrative Analysts who will oversee and manage the entire transformation program on a day-to-day basis. This group will be providing project management and oversight over the entire effort. Therefore they report directly to the Steering Committee.

ITSM Director

Remember that part of the overall ITSM solution includes an organizational component to manage it on an ongoing basis once it is in place. This is where to slot in the individual who will head your IT Service Management organization. Ideally it is recommended to get that individual on board at the start of the transformation program. This will allow for a much easier transition to production operations when things are ready. This shows where this individual links within the overall program.

Governance Track Leader

Governance Track Leadership will lead the team focusing on measurements, reporting and continuous service improvement cycle activities. The leadership itself should be tightly integrated with the ITSM program Manager and the Program Office. The Governance team itself is represented by Core, Extended and Advisor personnel farther down the organization structure.

ITSM Program Manager

This calls out the individual who will serve as overall Program Manager that is part of the Program Office. Strategically, this individual will become the ITSM Service Manager on an ongoing basis post transition. In a large global effort, this person will eventually report to the ITSM Director described earlier once implementation efforts have been completed. Smaller non-global efforts may decide to combine these two roles.

Regional ITSM Managers

Since this structure represents a global transition effort, regional ITSM Managers have been appointed to represent and oversee Program activities within key geographic areas of the company (i.e. Americas, Asia and Europe/Africa). They essentially serve as an extension of the ITSM Program Manager role with a focus on a geographic region.

Stakeholder Initiative Representatives

Here is where your Stakeholder Initiative teams will reside. They consist of personnel who are working on similar initiatives elsewhere in the enterprise. They serve in Extended, Advisor or Subject Expert team support roles depending on the engagement strategy selected to work with each initiative. Remember that Stakeholder Initiatives are other similar Service Management efforts taking place within your company (i.e. another Change Management effort taking place in another business unit).

Global Process Owners

Best practices for ITSM require a Process Owner for each ITSM process. During the transition effort, owners will coordinate and manage the overall design, objectives and operation of the processes assigned to them. They act as the single point of contact for all process implementation and Initial Win efforts associated with their process area. They also work with Process Owners of other ITSM processes to ensure a cohesive solution is built. In a global effort, these owners will lead and coordinate their process efforts from a worldwide perspective.

Local Process Owners

Since this structure represents a global transition effort, local process ownership has also been established at the regional levels (i.e. Americas, Europe, and Asia). A Local Process Owner will report to the Regional ITSM Manager that oversees the geography, but work directly with and for the Global Process Owners as part of the effort under a dotted line relationship.

ITSM Subject Matter Experts

This is where any ITSM consultants or other skilled personnel will be assisting and helping with many of the ITSM transition tasks. Note that these groups report to and work with the Global Process Owners as Subject Experts.

Organizational Track Leader

The Track Leader represents team personnel that will focus on the ITSM Organizational solution, training, business cultural change and awareness to meet the transition objectives. Members on this team will work across all ITSM processes working closely with each Process owner at the global level.

Extended Stakeholder Teams

This represents The Extended Team personnel as described earlier that will assist the Core Teams. Remember that each member in this group represents one or more business units. Each Extended team member is assigned to one or more Process, Technology, Organization or Governance Core Teams.

Core Teams

This represents the members that work day-to-day implementing the ITSM process, technology, organization, and governance solutions. Each member should be assigned to one of these teams and works within the corresponding Work Track for each assigned process.

Advisor Stakeholder Teams

This represents The Extended Team personnel as described earlier that will assist the Core Teams. Remember that each member in this group represents one or more business units. Each Extended team member is assigned to one or more Process, Technology, Organization or Governance Core Teams.

Coalition Teams

This entity may be necessary for very large or diverse transition efforts. It consists of groups of people that represent major geographical or organizational units within the enterprise that receive and execute the ITSM solutions once delivered. An example might be a business organization that has 200 data centers. The coalition team would then consist of selected personnel that represent groups of these centers (i.e. all centers in the U.S. or say 50 centers selected at random). This leverages the transition teams such that you would not have to add all 200 centers as individual stakeholders.

An example of a Coalition Team for global company with many operating centers might look like the following:

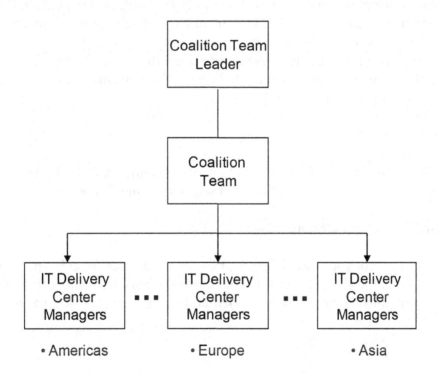

In the example above, the Coalition Team Leader acts that main single point of contact to the Implementation Program on behalf of all the delivery centers. The Team Leader helps assemble the Coalition Team, organizes Coalition Team meetings as necessary and ensures that delivery center input and participation is being adequately managed.

The Coalition Team consists of individuals that represent Delivery Center interests on behalf of a region or selected group of centers. They communicate issues and concerns for the centers they represent. They provide input to ITSM initiatives as they are developed from a Delivery Center perspective. They work with the centers they represent to ensure that participation and buy-in is maintained. In short, their role is very much like an Extended Team role.

The IT Delivery Center Manager can be any individual working within a local IT Center that has authority and responsibility for its operation. They are assigned to a Coalition Team Member and work closely with that person to raise issues and concerns from a local standpoint and provide feedback to the ITSM transition team.

The use of a Coalition Team organization plays an important role when transitioning to ITSM in a large organization with many operating centers. It greatly leverages the transition teams across many organizations to ensure that even the lowest level voice is heard on implementation decisions. You may pleasantly discover that there are some very good solutions and ideas that are operating in a particular center that might be worth using at all centers.

The Coalition Team also plays a critical role as ITSM solutions are deployed. It provides a mechanism for "pushing" solutions out to each of these centers in rapid fashion. Coalition Team members will be actively involved in ensuring that process solutions developed globally can actually be deployed within the units they represent. They work with the IT Delivery Center Managers to become the arms and legs for carrying ITSM solutions out to even the most smallest site.

Technology Track Leader

The Track Leader will lead the Technology team members in building the technical architecture and tools to support the ITSM solutions being developed.

Business Units and Other Organizations

Although not part of your teams directly, this entity has been included here to show the relationship that these groups have to your various teams. Remember that these will be directly represented through the Extended and Advisor Participation Roles discussed earlier.

Participation Roles

First and foremost, this effort is an organizational change effort. That is why the Participation Roles exist. They represent a way to maximize the solution buy-in across the enterprise. They allow for transition work to proceed efficiently minimizing the time needed to understand business requirements and agree solutions across a very large organization. If the burden for gathering those requirements and solution acceptance were placed solely with the Core Team members, the requirements gathering effort alone might take months or years. Not where we want to be!

Let's examine how this works. The goal is to leverage the construction of the ITSM solution by establishing four basic Participation Roles:

- Core Team

- Extended Team

- Advisor Team

- Subject Matter Expert

Every person involved with the implementation effort will have one of these roles. Each role is tied to a level of participation and type of responsibilities that people will bring to the overall effort. A description of each of these is as follows:

Core Team

This represents project team members who are doing heads down design, build and implementation of ITSM solutions. Members of this team are usually assigned full-time throughout the project. They are the hands-on people that will get ITSM solutions built and implemented.

Extended Team

This represents key business and IT stakeholders from all the business units in the enterprise who will be impacted by ITSM activities. They have two major responsibilities. First, they work closely with the Core Team to provide feedback on process, technical or organizational design decisions. Secondly, they represent the buy-in agreement for the business unit (or units) they represent.

This means that they carry key ITSM Program strategies back to their business unit, shop them around and obtain agreement within their unit. Alternatively, if all is not okay, they bring back concerns and issues to the Core Team. This greatly leverages the Core Team and allows the Core Team to focus on design and build activities versus getting sidetracked with having to go to each business unit in the enterprise to review what is being done. Members of this team usually are assigned part-time, but may have periods of time where more or less commitment is needed.

Advisor Team

This represents upper management personnel with a vested stake in the ITSM solution. Members from this team are used to provide final decisions on behalf of the business units they represent. They may be called on to assign an Extended Team member or Subject Matter Expert to assist the Core Team. They may simply desire to be notified of key implementation activities and events. Members of this team participate on an as-needed basis.

Subject Matter Expert

This represents a body of specific individuals who have unique knowledge or experience needed by the transition effort. They are used to bring in specialized expertise and usually provide knowledge and information on a consultative basis. Members of this team are assigned only when needed. They may serve in a full or part-time capacity at different points of the project.

Every implementation project team member and stakeholder is assigned to one Participation Role. While their role might change during the duration of the project, they will carry only one role at any given time. While the concept is fairly simple, do not be surprised how long it may take to pull it together. Experience has shown several challenges that you might run across within your organization:

- Business Units drag their feet for long periods of time looking for the "right" person to serve as an Extended or Advisor participant

- A Business Unit may provide a low level person as an Extended Team participant who is not really authorized to speak on behalf of that unit

- A Business Unit may provide a low level person as an Extended Team member and then replace them with a higher level manager part way through the implementation effort when they realize the kind of decisions being asked for.

As a result of the above, it is not unusual for much time to be wasted before the politics are sorted out. You can get ahead of this by being aware that this might happen and planning for it in advance. Another technique is to establish members in a "temporary" or interim role. This has allowed work to proceed without getting overly encumbered by politics and delays. Experience has shown that most of the time, the interim people chosen wind up seeing the effort through to the end anyway.

This approach has been used at many companies and has a proven track record of success when working with large organizations. It greatly serves to clarify implementation project roles and participation. Use of these teams provides a great way to obtain consensus and agreement across a large business organization very rapidly.

Implementation Roles

The purpose of identifying program implementation roles is to create blocks of related activities around the many implementation tasks. This makes the program organization much easier to manage and places each team member in easily recognizable functions.

In important fact to note about the exercise of identifying roles is that this is done totally independent of your current IT or business organization structure. The reason for this is to ensure that you stay focused on what is needed versus corrupting your thinking process with organizational issues and politics. These can be addressed by mapping roles into your organization structure as will be shown later in this chapter.

As a guideline, a list of implementation program roles is shown below. Note that these are recommendations. You may decide to have other roles depending on your specific implementation needs.

- Steering Group Member

- Program Manager

- Project Manager

- Process Owner

- Core Team Member

- Extended Team Stakeholder

- Advisor Team Stakeholder

- Subject Matter Expert

- Process Architect

- Tool Architect

- Tool Developer

- Organizational Change Leader

- Organizational Change Analyst

- Track Leader

- Facilitator

- Trainer

- Training Coordinator

- Technical Writer

- Coalition Team Leader

- Coalition Representative

- Administrative Analyst

The following pages list the above roles along with their definition and key activities. These are recommended roles only. Feel free to add, change or delete as you see fit. The purpose here is to give you a jump start on pulling your implementation organization together.

Steering Group Member

This role sets project direction, makes key decisions and provides final approval of Program Work Products. Key activities for this role include:

- Champions process solutions across the enterprise

- Conducts periodic meetings to monitor Program progress and issues

- Provides final review and approval of program deliverables

- Coordinates approvals from business units as necessary

- Identifies and appoints key Program team members

- Coordinates major program decisions that have been escalated to the Steering Group on a timely basis to meet program objectives

Program Manager

This role ensures Program Work Products are delivered on a correct and timely basis and ensures the objectives of the Transformation Program are met. This role has oversight over all implementation activities and manages and monitors the overall program effort. Key activities for this role include:

- Responsible for the overall project objectives.

- Provides direction to the project teams for work products due as well as the overall status of the project.

- Assigns Initial Win and Process Implementation projects to Project Managers.

- Provides status of work in progress and/or issues to the Executive Steering Committee

- Develops project work plan, schedule and staffing requirements.

- Communicates as required to executive management.

- Conducts weekly change, issues and status meetings to track progress and risks.

- Ensures that outstanding project management, process implementation and design requirements and/or issues are being addressed.

- Communicates activities and status of the overall Program throughout its lifecycle.

- Schedules workshops and meetings as required.

- Provides overall leadership and management of the overall Program.

- Coordinates activities of Project Managers and Project Office staff.

Project Manager

This role provides project management oversight and expertise to assist Core Teams in accomplishing their objectives. Project Managers work within the Program Office and are assigned to Initial Win and Process Implementation projects. Key activities for this role include:

- Responsible for assigned project objectives.

- Provides direction to the project teams for work products due as well as the overall status of the projects assigned.

- Co-ordinates activities with other project managers when necessary.

- Provides status of work in progress and/or issues to the Program Manager

- Develops project work plans, schedules and staffing requirements for projects.

- Communicates as required to executive management or Program Office staff.

- Conducts weekly change, issues and status meetings to track progress and risks with Core Teams assigned to.

- Ensures that outstanding project management, process implementation and design requirements and/or issues are being addressed for projects assigned.

- Escalates cross project issues or key management issues to the Program Manager.

- Communicates activities and status of each project assigned throughout its lifecycle.

- Schedules workshops and meetings as required.

- Provides overall leadership and management for the projects assigned.

Process Owner

This role ensures executive support of assigned ITSM processes, coordinates the various functions and work activities at all levels of a process, provides the authority or ability to make changes in the process as required, and manages assigned processes end-to-end so as to ensure optimal overall performance. Process Owners work with one another ensuring that process changes and improvements benefit the whole rather than help a specific function at the expense of another. Key activities for this role include:

- Responsible for the overall process objectives.

- Provides direction to the Core Teams for work.

- Monitors process maturity and progress throughout the implementation effort.

- Co-ordinates design decisions and activities with other Process Owners.

- Assists in development of project work plans, schedules and staffing requirements from a process perspective.

- Communicates as required to executive management and Program Office.

- Ensures that process implementation and design requirements are adequately identified and that process solution issues are being addressed.

- Identifies process and solution requirements to Technical Architecture Team.

- Identifies needed workshops and meetings as required to design and build process solutions.

- Coaches and teaches others about process concepts and solutions.

- Participates at communication events organized by the Organization Change Team.

- Provides overall leadership and management from a process perspective.

Core Team Member

This role provides heads down implementation of ITSM solutions. It also communicates with the Process Owners and Track Leaders to receive direction and to provide feedback on how well solutions are being implemented. Key activities for this role include:

- Assists in development of project work plans, schedules and staffing requirements

- Communicates with users of the process as to what is expected of them

- Assesses the current state of readiness and effort required to implement the processes, tools, governance and organization.

- Coaches the users of the process on tools and procedures.

- Communicates with the Process Owner on design, status and issues.

- Manages resources during detailed solution design and implementation.

- Ensures that documentation is developed and maintained.

- Participates at communication events organized by the Organization Change Team.

- Identifies additional resources as required to complete tasks such as writing procedures, developing job descriptions, producing analytical statistics or developing education material or architecting a technology solution.

- Ensures interfaces to other ITSM solution areas are working well.

Extended Team Stakeholder

This role actively participates in the development of ITSM work products and solutions. Responsible for representing the business or IT unit interests in the solutions being developed, managing communications between the Core Team and their department and obtaining departmental approval of process solutions being developed. Key activities for this role include:

- Actively assists in the review and development of Work Products.

- Provides input on solutions being developed to the Core Team.

- Coordinates decisions and feedback to the Core Team for key implementation and design decisions on behalf of the business units represented.

- Coordinates collection of solution requirements on behalf of the business units represented and feeds these to the Core Team.

- Obtains consensus and agreement on the solutions being developed from the business units represented.

Advisor Team Stakeholder

This role provides input and/or key decisions and recommendations to the Core Team on the solutions being implemented. It may also involve assigning others within the business unit represented to serve as Extended or Advisor Team members. Key activities for this role include:

- Reviews output of the implementation effort and provides feedback to Core Teams.

- Provides key decisions and approvals on a timely basis to meet implementation project needs.

- Assigns other department personnel to serve as additional Advisor and Extended team members as needed.

- Works in conjunction with other Advisor or Extended Team members within the department as needed.

Subject Matter Expert (SME)

This role provides expertise in technical, business, operational and/or managerial aspects for the design and implementation. Participation in the implementation is as required. This role may also provide specialized expertise in the design and implementation of process solutions as needed. Key activities for this role include:

- Provides technical, operational, business and/or managerial subject matter expertise.

- Provides input into the design of the procedures, tools or organization as required.

- Assists in the development of ITSM solutions by providing specialized expertise as required.

- Supports the development and execution of test scenarios designed to validate the functionality of the design.

- Validates the Design and Implementation Team designs for processes, tools and organization and any recommendations.

- Provides consultative and facilitation support to the Implementation Project Teams.

- Assists in creation of project work plans and implementation strategies.

- Provides Intellectual Capital as required during the Implementation Project.

- Coaches team members in specialized skill sets if required.

Process Architect

This role establishes the overall strategic process architecture and ensures a well-architected solution from a process perspective. The role provides consultative help around process modeling, design, build, implementation and rollout. One primary benefit and focus of this role is to coordinate common activities between the Process Core Teams to ensure maximum efficiency. Key activities for this role include:

- Coordinates all process activities across process Core Teams to ensure that ITSM solutions are integrated from a process perspective.

- Provides process expertise and input into the design, build and implementation of the ITSM solutions as required.

- Reviews process solutions under design for efficiencies in performance and resources to limit non-value labor and waste.

- May participate and lead in process modeling activities if used to design the ITSM solution.

- Supports the development and execution of test scenarios designed to validate the functionality of processes being designed and built.

- Supports governance activities from a process perspective.

- Coaches team members in specialized process skill sets if required.

Tool Architect

This role establishes the overall strategic tools architecture and ensures a well-architected set of technical solutions to support ITSM initiatives. The tool architect plays a dual role. It identifies and implements appropriate technology solutions that support the goals of the implementation. It also identifies new and changing technology solutions emerging in the marketplace that could provide value to the ITSM solutions being developed. This role also coordinates common technology related activities between all project teams involved in the implementation effort. Key activities for this role include:

- Ensures the tool architecture meets the strategic needs of the implementation effort.

- Coordinates technology product selections and tailoring.

- Supports cross team early launch planning from a technology perspective.

- Ensures maximum integration of tools.

- Coordinates technology product implementation activities.

- Coordinates technology customization and integration activities.

- Coordinates Technical resources to optimize use of technology solutions.

- Identifies ongoing support and maintenance for technologies chosen.

- Communicates chosen tool architectures and solutions to program teams.

- Interfaces to technology vendors as needed.

- Provides information on new and changing technologies to implementation teams.

Tool Developer

This role implements and customizes the technologies chosen to support the ITSM solutions being implemented. Activities also include provision of technical support during the implementation effort and assisting training activities by setting up training environments as necessary. Key activities for this role include:

- Understands the processes, tool requirements and data requirements for the technologies being implemented.

- Provides input to the detailed design for the processes based on technologies planned for implementation.

- Customizes implemented technologies based on the detailed design.

- Tests technologies for expected operation.

- Assists in developing procedures to install technologies.

- Assists training activities by installing and customizing the education technical environment.

- Implements technologies into test and production environments.

- Resolves problems with technologies.

- Provides technical support for technologies throughout the implementation effort.

Organizational Change Leader

This role develops and leads the organizational change effort to alter business culture and behaviors towards alignment with the solutions being implemented. It also serves to build the change communication strategy and develop the communications plan. It monitors and oversees all ITSM stakeholders and carefully crafts and controls all key messages about the ITSM effort, its progress, stated vision and goals. Key activities for this role include:

- Performs Stakeholder Management activities to identify Stakeholder concerns and issues with solutions being developed.

- Monitors stakeholder acceptance/rejection of solutions being developed.

- Crafts and controls key communications and messages about the implementation effort.

- Identifies opportunities to win acceptance of solutions being developed by those who are impacted.

- Identifies channels for communications and builds the overall communications plan.

- Develops a Resistance Management Plan to provide strategies for dealing with rejection or resistance to solutions being developed.

- Ensures appropriate levels of the organization are involved and demonstrating active commitment and leadership to the solutions being developed.

- Coaching Senior Management and other key personnel to help them "walk the talk" and demonstrate commitment to the ITSM solution.

Organizational Change Analyst

This role supports the Organizational Change Leader with a variety of administrative and organizational change development tasks as needed to meet the goals of the ITSM transition effort. Key activities for this role include:

- Maintains stakeholder documentation as needed.

- Develops training and presentation materials.

- Schedules training for ITSM team members as needed.

- Prepares artifacts related to the ITSM communications strategy such as newsletters, program giveaways and communication reports.

- Maintains e-mail distribution lists for stakeholders and ITSM implementation Program personnel.

- Schedules key meetings with stakeholder teams and steering group members.

- Takes and publishes notes at key program meetings and workshops that involve stakeholders.

- Other administrative tasks as needed to support the transition effort.

Track Leader

This role provides leadership to a Program Track (i.e. Process, Technology, Organization and Governance). It is responsible for all activities within that track and coordinates track activities with other Track Leaders. It provides a single point of contact to the Program Office and Process Owners for coordinating Work Tack Tasks. Key activities for this role include:

- Manages all activities that occur within the track assigned.

- Single point of contact for all Track Activities and issue resolution.

- Coordinates track activities with peer Track Leaders.

- Develops work plans for all track activities.

- Reviews Work Products produced by track and coordinates their acceptance with the Program Office.

- Works to resolve track related issues.

- Escalation point for track team members to resolve track issues or issues that need coordination across multiple tracks.

- Coordinates status reporting for track activities.

Facilitator

This role leads and conducts working sessions and meetings in a neutral fashion to ensure that the goals of those sessions and meetings are met. Key activities for this role include:

- Leading meetings and working sessions in a neutral manner to ensure goals and outcomes of those sessions are being met.

- Developing session detail agendas and agrees these with those involved.

- Developing discussion strategies and methods to ensure all participants are involved and to obtain consensus on key decisions in an efficient manner.

- Monitoring sessions to make sure all sides of discussed issues are being considered and that session groups do not "go with the flow" unless truly in agreement.

- Identifying needed materials and supplies for meetings.

- Providing feedback to Organization Change Team on participant acceptance of meeting issues and activities based on observation during meetings.

Trainer

This role provides training for process procedures, governance policies and use of processes and tools. It identifies training needs and requirements. It builds the needed curriculum paths for each ITSM implementation team member and stakeholder. It leads and conducts training sessions. It also leads in the development of training materials. Training scope includes the Implementation Program Teams as well as business and IT staff that will utilize the ITSM solutions. Key activities for this role include:

- Identify needed training and curriculum paths for ITSM implementation team members and business stakeholders.

- Training users on processes, use of tools and procedures.

- Leading development of training material as needed.

- Leading and conducting training sessions

- Designing and building training curriculum for implementation personnel and business units impacted by ITSM services

- Identifying ITSM certification needs and requirements

- Coordinating Subject Matter Experts to assist with training as needed.

- Aligning training curriculum and events with Organization Change activities and plans.

- Publicizing training events and activities.

Training Coordinator

This role provides administration support for training activities. It assists with preparation of training material, manages training schedules, training registration and attendance. It tracks attendance at training and monitors status of training for each implementation team member and business stakeholder. Key activities for this role include:

- Preparing training material as needed.

- Administering ITSM certifications and tracking certification results.

- Tracking attendance at training sessions and training progress for implementation personnel and business stakeholders.

- Administering training calendar and schedules.

- Handling administrative tasks associated with vendor provided training.

- Registering personnel for training activities and events.

- Other administrative tasks as directed by Trainers.

Technical Writer

This role sets standards for how processes and procedures are to be documented. It documents process guides and work instructions in a manner that is easily understood by those executing the processes. It participates in the documentation of tool architectures and tool changes. It builds and publishes templates for presentations and key ITSM forms. Key activities for this role include:

- Setting standards for how processes and procedures should be documented.

- Produces documentation for process guides and procedures.

- Provides consulting guidance on how to best present documented information so it is quickly and easily understood.

- Identifies improvements for existing documentation.

- Designs and builds templates for key presentations and process work products.

- Designs and builds templates for forms used as part of the ITSM solution.

Coalition Team Leader

This role leads and organizes activities and meetings with Coalition Team members for larger ITSM transition efforts with many IT service organizations and delivery centers. It serves to ensure that ITSM solutions being developed will be able to be implemented across all the organizations represented. It also plays a main part in rolling out ITSM solutions to those organizations. It organizes coalition teams and leads coalition team activities. It also collects and summarizes input from Coalition team members for ITSM transition teams. This role serves as an Extended Team Stakeholder. Key activities for this role include:

- Organizing and leading Coalition Team meetings.

- Identifying the appropriate Coalition Team membership needed to adequately represent all the organizations it was established for.

- Identifying and obtaining Coalition Team members.

- Collecting and summarizing Coalition Team input and feedback on ITSM design decisions and solutions.

- Leading rollout efforts on behalf of the organizations to receive and operate ITSM solutions once they are built.

- Assisting ITSM solution design efforts by summarizing key ITSM related solutions that may be operating currently at the organizations represented.

Coalition Representative

This role provides a single point of contact into one or more IT organizations and service delivery centers. It represents the concerns and ideas of those organizations. It provides input and feedback to ITSM solution designs and plans based on feasibility within the current infrastructure, operations and culture with the organizations represented. It identifies ITSM related solutions that may be operating in some of the organizations represented that may be of help to those designing and building ITSM services. It also assists with rollout and implementation of ITSM agreed solutions at the organizations represented. This role works with and reports to the Coalition Team leader. Key activities for this role include:

- Reviewing ITSM plans designs and key decisions with IT and business staff at the organizations represented.

- Providing feedback, concerns and issues that are raised by represented organizations to the Coalition Team Leader on ITSM plans and designs.

- Identifying ITSM related solutions already in place and operating that may be of help to ITSM implementation teams.

- Assisting in the development of rollout plans unique to organizations represented.

- Rolling out ITSM agreed solutions to the organizations represented.

- Attending Coalition Team meetings.

Administrative Analyst

This role performs administrative and clerical duties and activities as needed to support the ITSM Program. This role mainly resides within the Program Office but may also exist within other teams as needed. Key activities for this role include:

- Gathering and collating Program status report information.

- Administering Program document repositories and web pages.

- Collecting labor hour/time reporting information from Program participants.

- Managing Program Email distribution lists.

- Managing and publishing the Program calendar.

- Setting up Program Meetings and schedules.

- Coordinating travel arrangements for Program participants.

- Other duties as directed by the Program Office or other authorized team members.

Pulling the Implementation Teams Together

The previous sections in this chapter addressed various organizational elements that need to be in place for executing on an ITSM transition effort. How should all of this come together? A structured approach is presented here for doing this. It consists of the following seven steps:

1. Identify the Implementation Roles needed in the program to execute transformation tasks.

2. Identify the key program and transition activities that each role will handle

3. Identify the skills and skill levels embodied in each role

4. Map Implementation Roles into an organization structure that will be used to perform the transition effort

5. Identify transition team members and match them with one or more roles

6. Identify the Participation Roles for each team member

7. Place team members into the organization structure

The above series of steps will help you with building a well-defined transition team structure with clearly defined roles and responsibilities in a structured manner. Let's now look at each of these steps in more detail.

Identifying the Program Implementation Roles (Step 1)

In this step you will identify a list of Implementation Roles that you think you will need. The list provided earlier in this chapter is good start. Overall, you should not end up with more than 20-30 roles for a large implementation with some number less if your effort is smaller. Think about the tasks that need to be done and look at the Work Breakdown plans and structures for guidance. Think about what roles might be necessary for what needs to get done.

Identifying the Program Implementation Role Activities (Step 2)

Once the Implementation Roles have been defined, the next step is to identify a set of related key activities that belong to each role. These are sets of related kinds of tasks that are the responsibility of that role. As an example, the Process Owner role might have the following key activities associated with it:

*   Responsible for the overall process objectives.

*   Provides direction to the process Core Teams for work.

*   Monitors process maturity and progress throughout the transition effort.

- Co-ordinates design decisions and activities with other Process Owners.

- Assists in development of project work plans, schedules and staffing requirements from a process perspective.

- Etc . . . etc . . .

As a way to kick start your efforts, you can refer back to the Program Implementation Roles section of this chapter and review the activities described there.

Identifying the Skills and Skill Levels (Step 3)

For every role, the ability to execute the key activities associated with it will require certain skills. These are described in two ways: a set of skill traits that belong to that role and a skill level indicator for each trait that identifies the depth of expertise needed for that role.

A skill trait is a label that describes a distinguishing proficiency, talent or ability that will be required by the person selected to be in a specific role. A skill level would refer to indicator of how much of that skill might be needed for that role. For example: the role of Program Manager might require a Negotiation skill trait and the person filling that role needs to be at a level 5 (Very High) in that skill trait for that role. As another example, a Skill Level of 1 attached to ITSM would indicate that only an awareness of ITSM is needed for a particular role.

You can begin to see where this heading. These will be used to help identify and screen appropriate personnel in your organization that will serve in the transition program. This will allow you to establish fairly clear criteria on the kinds and depth of skills that you will be looking for on your teams.

A set of example skill traits is listed below. Feel free to modify this to suit your needs. For example, you may wish to add a specific tool set expertise or not use everything shown on this list.

Example Skill Traits:

- Customer Relationship
- Negotiation
- Project Management
- Technical Architecture
- Process Architecture
- Business Skills
- Communications
- Leadership
- Writing
- Teaching/Coaching
- ITSM

- Administrative
- Analytical
- Political/Social
- Planning
- Operational Expertise

To handle skill levels a simple skills rating system from 0 to 5 and related criteria can be used. You may wish to alter this based on the general practices of your organization and your specific program needs. An example set of skill levels and criteria are as follows:

Skill Level	Criteria
0	Skill not necessary
1	Awareness of what has to be done
2	Basic level but needs supervision
3	Can perform many tasks without supervision
4	Can work independently
5	Can lead others

Mapping Roles into the Organization Structure (Step 4)

Now that needed roles have been identified along with skills and skill levels, the next step is to map your defined roles into a program organization structure. The first step is to identify how your transition organization structure will look. What entities will exist in it? Which entities are managed or controlled by other entities?

To build the organization structure, you can use the ITSM Transition Organization structure shown in this chapter and the various structures described individually for each ITSM Work Stage chapter. Then modify these to fit your specific needs. Next, assign each role by name into the organization position where it makes sense. For example, the role of Project Manager might be assigned to the Program Office organizational entity in your organizational structure.

Note that it is okay for some organization positions to have more than one role assigned to it. For example, the Program Office organization entity could many roles assigned to it such as *Program Manager, Project Manager* and *Administrative Analyst*.

Every entity in your organizational structure should have at least one role assigned to it. If by some chance you stumble across an organizational entity without any roles, then you either do not need it in the first place, or you need to go back and define an additional role and set of skills for it.

As a starting point, the Program Role Table shown on the following page illustrates a possible mapping of recommended roles to organization entities. Since your implementation organization structure may differ, the table maps the roles described in this chapter to the transition structure described in this chapter. When looking at the Program Role column, remember that these are Program Roles, not organization or job titles.

Role Mapping Example

Program Role	Organization Structure Entity
Steering Group Member	Global ITSM Steering Committee
Program Manager	Global ITSM Program Office ITSM Program Manager, ITSM Director Regional ITSM Manager
Project Manager	Global ITSM Program Office
Process Owner	Global Process Owner Local Process Owner
Core Team Member	Core Team
Extended Team Stakeholder	Extended Stakeholder Team Stakeholder Initiative Representative
Advisor Team Stakeholder	Advisor Stakeholder Team Stakeholder Initiative Representative
Subject Matter Expert	ITSM Subject Matter Expert Stakeholder Initiative Representative Technology Team Organization Team Governance Team
Process Architect	ITSM Subject Matter Expert
Tool Architect	Technology Team
Tool Developer	Technology Team
Organizational Change Leader	Organizational Team
Organizational Change Analyst	Organizational Team
Track Leader	Process Team Organization Team Technology Team Governance Team
Facilitator	Organizational Team
Trainer	Organizational Team
Training Coordinator	Organizational Team
Technical Writer	Process Team Organization Team Technology Team Governance Team
Coalition Team Leader	Coalition Team
Coalition Representative	Coalition Team
Administrative Analyst	Global ITSM Program Office

Matching Team members To Roles (Step 5)

At this point, there is enough information about the program roles and skills needed to identify personnel who will be assigned to them. This is just a matter of selecting personnel who will work in the Program and assigning one or more roles to them. Here you will use the skills traits and skill levels to help screen and place which individuals might fit into which roles.

In a very large effort, it is possible that a role will be a full-time effort for one or more staffing resources. A much smaller effort may combine multiple roles with a single individual. Several things need to be considered when assigning staff to the transition program:

- Selected resources must match the roles and skill levels needed.

- Selected resources must be available to work on the effort.

- Selected resources are committed for the durations needed to accomplish Program objectives.

- Ideally, selected resources are excited by the effort and show a willingness to serve on the transition team.

Resources serving on various teams typically are chosen based on the knowledge or political sway they hold in the organization. Experience with many ITSM transitions has shown these teams to be fairly fluid in terms of commitment. This has usually occurred for the following reasons:

- Subject Matter Experts are usually only needed for spot expertise and will come in and out of the effort over time as needed.

- Business Units may initially select someone based solely on availability to serve as an Extended Team member. As the program progresses, it becomes evident that the person chosen may be the wrong person to truly represent that unit. The person then gets replaced by another individual deemed better suited for that role.

- Business Units may change their position on who truly "represents our department" midway into the effort and add or replace Extended or Advisor team members.

- New stakeholders may be discovered midway through the effort. These then are added over time throughout the program lifecycle. This is okay. If you work in a large organization, the time spent researching for every possible stakeholder would be enormous and your effort may never have gotten started.

- Business Unit political issues introduce delays and changes over who should really be assigned to the Program.

These observations are pretty common when going through this kind of an effort. During the Planning Work Stage, you may wish to identify these as risks and develop a mitigation plan to deal with them if they occur. The largest impact is usually found in bringing new members up to speed midway in the effort and making sure the rest of the transition team is aware of the changes.

Identify Participation Role for Each Member (Step 6)

For this step, you will be identifying which Participation Role each member of the program team will play. This means tagging each person on the team as a Core, Extended, Advisor or Subject Matter Expert. Go back through the Work Breakdown Structure and determine who will be doing which tasks. Will these require full time or part time participation?

The following guidelines may help you decide who might fall into what category:

- If full-time participation is required, you may wish to assign a *Core Team* role

- If part-time, but directly responsible for developing a Work Product you may wish to assign a *Core Team* role

- If person is in Senior Management or other key decision/approval making capacity, assign them an *Advisor Team* role

- If person is on team because they represent a business unit, other IT department or another related initiative, assign them an *Extended Team* role

- If person is on the team because they have a needed skill or expertise, assign them a *Subject Matter Expert* role

Placing Team Members into Organization Structure (Step 7)

Now that program candidates have been selected and mapped into roles, the last step is to place them officially within the Program Organization structure determined earlier. In the course of this exercise, you may find that you are making slight modifications to the Organization Structure itself to accommodate existing reporting relationships, politics or other organizational issues.

At this point, you should have a fairly well defined program transition team. Program roles and general responsibilities should be well-defined and there should be almost no questions by anyone on the effort where they will be participating as the program moves forward.

The approach recommended here is a good structured way of pulling your program organization together. More importantly, you will find that a similar approach will be used in addressing the organization that will be needed to operate your ITSM solutions on an ongoing basis once they are in place.

Chapter 6

ITSM Visioning Work Stage

> "However beautiful the strategy, you should occasionally look at the results."
> —Winston Churchill

A Program operating without a vision is basically operating blind.

Establishment of a solid ITSM Vision is one of the harder Work Stages to execute for people doing ITSM transitions. If done well, you will find the effort was well worth it. Resist temptation to jump into an ITSM Assessment or transition effort without doing this first. This is the stage where you will establish a solid case for moving to an ITSM solution, getting key decision makers on board with the effort and stirring excitement over the benefits. The goal of this chapter is to provide you with a structured approach for accomplishing all this.

You cannot blindly go jump in assessing and designing processes right away. There must be a focus and understanding of what the business is looking for. Without this, you may miss the real benefits sought after by your company let alone waste time and money.

This becomes even more critical if you will be using outside consulting resources to conduct the initial assessment efforts. Outsiders have little or no idea of your company's issues and problems. You will be paying expensive dollars to them for basic research to figure this out. Worse yet, they may produce assessment findings that show gaps against the ITSM standard, but cannot articulate why those things should be fixed now or which items will provide the greatest business benefit.

The main goal of the ITSM Visioning Work Stage is to identify and agree the vision, scope and key business benefits to be achieved by the ITSM effort. This task is absolutely critical and should be performed before any transition work gets started. If this is neglected or done poorly, there is a major risk that the effort itself will stall and die later down the road. Management will perceive that business issues are not being addressed and staff may start asking why the effort is necessary and drift back to old practices and habits.

Therefore, the key goal is to set the stage for why this effort is important in the first place. At the end of this Work Stage, your organization should have a solid view of the benefits and problems to be solved. There should be enough agreement among key stakeholders and sponsors that things should proceed to the next step.

ITSM Visioning Work Stage Approach

If we look at the Visioning effort over an approximate 4 week timeline, the key Work Tasks might be timed to look like the following:

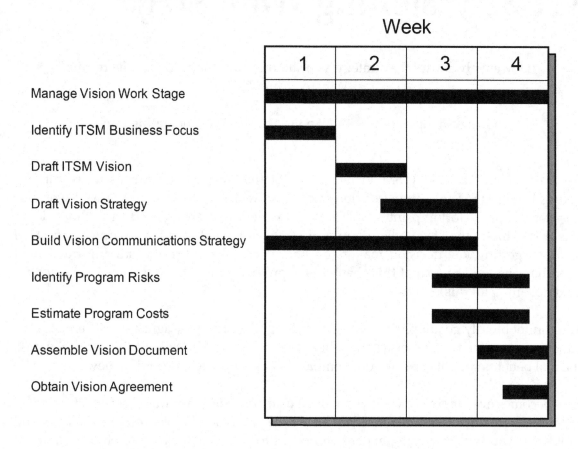

A detailed Work Breakdown Structure is depicted on the following page:

ITSM Visioning Work Stage—Work Breakdown Table

R=Responsible
A=Assists
C=Consults
I-Informed

Work Task	Work Step	Org	Proc	Tech	Gov	Mgt	Work Product
Manage Vision Work Stage	Confirm sponsor/champion	C	C	I	I	R	
	Assemble Core Team	A	C	C	C	R	
	Assemble Extended Team	A	C	C	C	R	
	Assemble Advisor Team	A	C	C	C	R	
	Finalize ITSM Visioning Team List	A	C	C	C	R	ITSM Vision Team List
	Manage Vision Stage Activities	A	A	A	A	R	
	Report Vision Stage Status	A	A	A	A	R	Status Reports
	Schedule Vision Stage Meetings	A	A	A	A	R	
	Inventory existing IT services	I	R	C	I	I	Inventory of IT services
	Inventory ITSM technologies	I	C	R	I	I	Inventory of Technologies
	Identify customers and groups	R	C	C	I	I	Inventory of customers
	Identify current organization structure	R	C	C	C	I	
Identify ITSM Business Focus	Conduct Customer Needs Survey	R	C	C	C	I	Customer Needs Survey Results
	Conduct ITSM Focus Survey	R	C	C	C	I	ITSM Focus Survey Results
	Conduct ITSM Benefits Survey	R	C	C	C	I	ITSM Benefits Survey Results
	Conduct ITSM Cultural Survey	R	C	C	C	I	ITSM Cultural Survey Results
	Conduct ITSM Labor Survey	R	C	C	C	I	ITSM Labor Survey Results
	Draft Business Focus Summary	R	C	C	C	I	Business Focus Summary
	Schedule and conduct Vision Stage Kickoff Session	R	A	A	A	I	Kickoff Session Agenda

Work Task	Work Step	Org	Proc	Tech	Gov	Mgt	Work Product
Draft ITSM Vision	Prepare for visioning workshop	R	A	A	A	A	Visioning Workshop Agenda
	Conduct IT Values Brainstorm	R	A	A	A	I	IT Values Analysis Results
	Conduct Vision Statement Brainstorm	R	A	A	A	I	
	Conduct Vision What If-Up Activity	R	A	A	A	I	
	Conduct Vision SWOT Analysis	R	A	A	A	I	SWOT Analysis Results
	Identify overall Guiding Principles	A	R	A	A	I	Overall Guiding Principles
	Draft Vision Statement	A	R	A	A	I	
	Identify desired future state impacts	A	R	A	A	I	Future State Impacts
	Confirm Vision Statement	R	A	A	A	A	Vision Statement Draft
Draft Vision Strategy	Review survey results and Vision	C	R	C	I	I	
	Review IT needs from survey results	C	R	A	A	I	
	Perform ITSM QFD Analysis	A	R	A	C	I	ITSM QFD Analysis
	Summarize and interpret QFD results	A	R	A	C	I	
	Identify ITSM Management Model	A	R	A	C	I	ITSM Management Model
	Draft service demand factors	C	R	A	I	I	Service Demand Factors
	Identify Overall Vision Strategy	A	R	A	A	A	Vision Strategy Draft
Build Vision Communications Strategy	Draft impact analysis	R	A	A	A	I	Impact Analysis
	Draft stakeholder analysis (WIIFM)	R	C	C	C	I	Stakeholder Analysis
	Draft vision benefits analysis	R	A	C	C	I	Inventory of Vision Benefits
	Draft program success metrics	A	A	A	R	I	Program Success Metrics
	Build vision communications strategy	R	C	C	C	C	Vision Communication Strategy
Identify Program Risks	Identify key delivery issues	R	A	A	C	I	Inventory of Vision Issues
	Identify cultural concerns from IT Service Management Cultural Survey	R	C	C	C	I	Inventory of Cultural Risks
	Draft program risk analysis	A	A	A	R	C	Program Risk Analysis
	Identify risk mitigation actions	A	A	A	R	C	Risk Mitigation Actions

Work Task	Work Step	Org	Proc	Tech	Gov	Mgt	Work Product
Estimate Program Costs	Draft Assessment Work Stage Plan	A	A	A	A	R	Assessment Work Stage Plan
	Identify cost factors	A	A	A	R	A	Inventory of Cost Factors
	Document cost assumptions and calculations	A	A	A	R	A	
	Draft estimated high level costs	A	A	A	R	A	Estimated Program Costs
	Build preliminary cost model	C	A	A	R	C	Preliminary Cost Model
Assemble Vision Document	Review Vision Work Stage Work Products	R	C	C	C	I	
	Draft vision document	A	A	A	R	I	Vision Document Draft
	Draft vision presentations	R	A	A	A	C	Vision Presentations
	Review Vision document with stakeholders and Core Team members	R	C	C	C	I	Vision Feedback and Updates
	Update Vision document based on feedback	R	C	C	C	I	Vision Document
	Review vision with key decision makers	A	A	A	A	R	
	Obtain consensus on vision	R	A	A	A	A	Vision Agreement
Obtain Vision Agreement	Assemble Program Charter Draft	A	A	A	A	R	Program Charter Draft
	Assemble Program Steering Group	A	C	C	C	R	Steering Group Member List
	Identify ITSM Assessment Work Stage Team candidates	A	A	A	A	R	ITSM Assessment Team Candidates List
	Execute Vision Communications Strategy	R	A	A	A	A	
	Obtain Go-Forward Agreement	A	A	A	A	R	Go-Forward Agreement

Visioning Key Work Products

The table summarizes key Work Products produced by the ITSM Visioning Stage. Other Work Products are described further in the ITSM Vision Work Task Considerations section of this chapter.

ITSM Visioning Key Work Products

Work Product	Description
ITSM Vision Team List	Lists personnel confirmed to be working on the ITSM Vision phase Identifies members by Core, Extended, Advisor and Subject matter affiliations Shows contact information
Business Focus Summary	Summarizes results from Customer Needs, ITSM Focus, ITSM benefits, ITSM Labor and ITSM Cultural Survey Identifies candidate ITSM Processes that may need more attention than others
Vision Statement Draft	Brief, rallying statement of future state that serves to stimulate excitement and interest in what is to be achieved with the ITSM Program Supported by IT values, guiding principles and high level impacts of what the ITSM Vision will mean if implemented
Vision Strategy Draft	Documents high level view of how the Vision will be put into place Identifies the transition from current state to the future state Identifies core types of technologies that will support the Vision Identifies service demand factors that impact the Vision Identifies how the Vision will be managed and governed at a high level
Vision Communications Strategy	Documents strategy and plans for communicating the ITSM Vision throughout the business organization Identifies key channels for communications (i.e. Brown-Bag sessions, E-Mails, Announcements Identifies key timeframes and milestones for communication activities Identifies roles and responsibilities for communications activities
Program Risk Analysis	Identifies high level program risks (examples might include costs, staffing levels, cultural changes, skill gaps, etc.) Identifies planned mitigation strategies to overcome the risks identified

Work Product	Description
Estimated Program Costs	High level estimate of cost ranges for the ITSM Program Identifies cost factors and assumptions Documents a preliminary cost model Shows more detailed costs for the ITSM Assessment Work Stage effort
Vision Document	Reiterates the Vision statement Communicates key benefits specific to the business organization (Why is this important?) Communicates sense of urgency (what if nothing is done? Why should this be done now?) Identifies general scope, stakeholders and impacts (Users, Customers, etc.) Communicates benefits for each type of stakeholder (What's in it for them?) Communicates general timeframes for when key results might be expected Lists program success factors Lists high level demand factors that will influence ITSM services Identifies potential program risks and mitigations May include high level cost estimates Indicates a high level view of how the vision will be achieved (i.e. use ITSM, Six Sigma, etc.) May be variations on this for different audiences
Program Charter Draft	Structured per corporate guidelines to establish program and obtain funding Identifies program goals and objectives Indicates estimated program costs Identifies program key objectives and how they will be measured Identifies duration of the program
Go-Forward Agreement	Documented management approval to proceed to the next phase of effort Approved budget for the next phase of effort Identified staffing team for next phase of effort (This should include Process Owners) Identified consultants if these will be used for the next phase of effort

ITSM Visioning Implementation Team Organization

The recommended ITSM Visioning Implementation Team Organization is that pictured below:

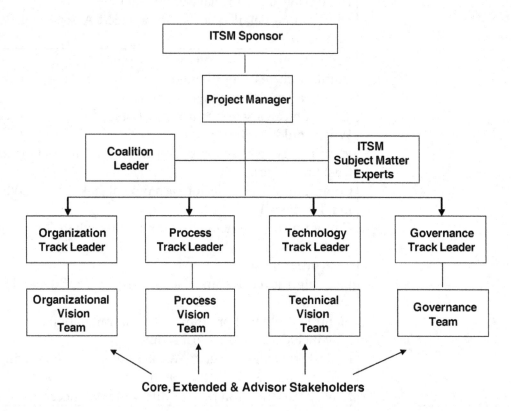

Drawing from the earlier Transformation Organization chapter, the Visioning team organization will be a smaller subset. Note that the structure allows for the four Work Tracks: Organization, Process, Technology and Governance. The Management Work Track is covered by the Project Manager.

The teams that work within each track should consist of Core, Extended and Stakeholder types of individuals as described earlier. While the selection of the team members is not as large or broad as the later implementation stages, a solid representation of key decision makers should be included.

ITSM Visioning Work Task Considerations

This section describes each of the ITSM Visioning Work Tasks in more detail. Practical tips and ideas are shown here for each Work Task that you may find useful. Feel free to steal any of the ideas presented here or modify them to suit your specific needs.

Manage Vision Work Stage Task

This is not unlike general project management. The Project manager will do all the typical tasks of setting up the project, time reporting, obtaining work space and meeting rooms, setting up status reporting and other administrative functions to ensure this phase operates smoothly. Most importantly, this Work Task will ensure that the Visioning Team is appropriately staffed and that people are ready to work on it. Much effort will be needed to make sure the right people are available at needed times.

Identify ITSM Business Focus Task

You may be confused as to why this Work Task is included. After all, did not management approve this effort already? Did they not budget for it? Did they not state that they strongly believe that ITSM is important before all this started?

There is an interesting thing about Senior Management when it comes to ITSM. Many have positive views about what it might do for their company. Few really understand what it takes to transition and operate to an ITSM state. There may be many underlying perceptions and assumptions about which business problems ITSM will fix. For example, a senior manager drowning under constant service outages might be thinking this will fix many of those problems.

The purpose of this task is to draw all those preconceptions and assumptions out on the table where everyone can see them. This is very critical since you want to make sure that there is consensus on the business issues and problems to be addressed by the implementation effort. If you embark on the implementation effort without this understanding, you will be dead in the water halfway through because Senior Management had different expectations about what they would be getting.

There are several tools that you can use to draw out expectations. These are described in the ITSM Visioning Tools and Aids section of this chapter and can be found in the ITSMLib ITM Implementation Library. The tools are:

Customer Needs Survey

This is a simple survey that can be used to identify what your customers really want out of their IT Support and Delivery Services.

IT Service Management Focus Survey

This is a simple survey that can be used to identify why IT staff and management desire to look at ITSM as a solution. It should be given to IT management and staff.

IT Service Management Benefits Survey

This is a simple survey that can be used to identify what benefits people are hoping for as the result of implementing ITSM. This should be given to IT management and staff as well as non-IT Business Units.

IT Service Management Cultural Survey

This is a simple survey that can be used to identify how people view the current IT service delivery culture as it exists today. The output of this will be used later on in this Work Stage to identify cultural risks and barriers that may threaten your implementation efforts. This should be given separately to IT Management, IT Staff and non-IT Business Units. You may find some interesting differences in how each group responds to this.

IT Service Management Labor Survey

This is a simple survey that can be used to identify the general labor mix of where people are spending their time. This should be given to IT service support and delivery staff. High percentages in non-value and overhead labor areas may indicate opportunities where ITSM can make a difference.

The emphasis on the above surveys is simplicity. Each survey should take not much longer than 10 minutes to complete. They are quick and easy to use but provide lots of information about the current climate for change within your company. That is why they are included here. The output of the surveys will be used as background to help drive the Vision Statement, Approach and Risks.

Draft ITSM Vision Task

In this Work Task, the actual ITSM Vision Statement will be drafted. The Vision statement is extremely important. This will be the flag that everyone will rally around that serves to stimulate excitement and interest in what is to be achieved. It does not need to be too detailed. If those reading or hearing it become excited and ask "tell me more . . ." it will have served its purpose.

Several rules on Vision Statements need to be considered:

- They should usually be one sentence

- They should identify a stated goal in generic terms

- They should not be cluttered with statements about how the Vision will be achieved

- They should not include specific product names or standards

Here are some good examples of ITSM Vision statements that have been used at other companies:

We will set the industry standard for how to operate a global IT infrastructure that is admired around the world—most importantly—by our customers as a reason to buy products and services from our company.

We support our business goals of transformation into a leading company by providing the highest quality IT service infrastructure that delivers the most reliable, accurate and secure information services in the world.

We will always protect our corporate investments in IT solutions and infrastructure by ensuring that they will be **deployable and operable** on a day-to-day basis at acceptable cost.

We will deliver a range of responsive and integrated services that will meet, and where appropriate, exceed our customers' expectations.

We will set the standard for how to operate a set of infrastructure management services that is admired around the world—most importantly—by our associates who will proudly use our services

We will create and operate a set of infrastructure management services parallel to no one else in our industry

Here are some not so good examples of ITSM Vision statements that have been used at other companies:

We will use ITSM to drive the standard for how our IT services will be delivered.

We will achieve an IT Service Management CMM level of 4.0 or higher.

Through improved ITSM alignment and process maturity, we will be able to:

- Support and further corporate mandates and objectives

- Effectively utilize resources through the clear definition of roles and responsibilities

- Enable us to absorb growth with a virtually flat staffing level by way of achieving process excellence

- Align service delivery with service level requirements

- Implement changes faster, improving business agility

- Reduce the volume of Incidents resulting from Change

- Improve capture and utilization of operational data for proactive purposes

- Provide efficient and effective response to service outages

- Reduce the number of service outages

- Resolve issues faster, minimizing business impact

- Analyze and intercept Incidents and Problems before they impact our customers

In the above examples, the first two really address how a vision will be achieved, but does not state what the vision is. In short, those companies are building a road, but have not identified where it will lead to. The last example, which is fairly typical, states some good things, but is far too complicated and long to describe to someone in a 30 second elevator speech. Just reading it may cause some eyes to glaze over.

The following questions may be used to help you think about how to set some possible vision statements:

- Who are our customers?

- How should we appear to our customers?

- How should we appear to our stakeholders?

- What must we excel at with our customers and stakeholders?

- How will we sustain our ability to change and improve?

Draft Vision Strategy Task

The Vision Strategy will identify the general direction and approach to achieve the Vision Statement developed earlier. Here you will be defining the key actions and standards that will be undertaken to achieve the Vision. In short, you will be answering the question "How will we achieve the Vision?"

The strategy should consist of a number of areas which you may wish to consider. These are described as follows:

Approach

At this early stage, you will want to identify the key actions and general direction that will be taken to support the Vision. Statements at a broad level that one might make at this stage might include statements such as:

- We will utilize a process driven approach based on the industry standard ITSM Process Framework as a way to organize how we will deliver our IT Services.

- We will embark on a Service Improvement Program starting with an initial assessment of our service delivery and service support capabilities.

- We will review our supporting tool sets and how they have been implemented to make sure they fully support our service support and service delivery strategy.

You can leverage results of the surveys done previously to drive the elements of your approach. These will help identify the IT values that exist within your company and the business focus areas that are of concern.

Overall Guiding Principles

Guiding Principles are statements about how IT Services should operate. They support the IT Service Management Vision by existing as critical statements of direction that will have major impact on how IT services should be designed and operated. They are derived from your company's basic beliefs, experience, priorities, values and underlying culture.

An inventory of Overall Guiding Principles is provided in a separate chapter in this book. You will see that there are many there to choose from, but the actual number you use will most likely be a small subset. Many more of these can be found in the ITSMLib ITSM Implementation Library as Design Guidelines.

Transition Approach

This describes a high level view of how the organization will transition from its current state to the future state as outlined by the Vision. Much of the ITSM transition approach, as described in this book, can be drawn upon to describe how your organization will proceed. This should be summarized at a very high level. For example, the overall transition approach showing what Work Stages will occur can be used without getting into much more detail.

Business Alignment

To align the strategy specific to your business organization, the ITSM Alignment tool described in the ITSM Visioning Tools and Aids section of this chapter and also available on ITSMLib. This ensures that the effort will be focused on what is important to your business. The tool will serve to help prioritize which aspects of ITSM need to be addressed more than others.

The ITSM Alignment Tool is a recommended way to determine where your efforts will get the biggest bang for the buck. This tool is loosely based on the Quality Function Deployment (QFD) approach used with Six-Sigma. The tool will allow you to work through which ITSM areas will be most important based on customer needs versus solely what IT may think needs to be done.

Management Model

This describes a high level view that shows what key management functions will be in the future state vision and how they interact. The description of this will be done by building a management model that shows the key components and their overall flow.

The goal is to illustrate, at very high level, how the service vision will look and interact with your company's organization and culture. Two examples of these are presented on the following page. In both models, for example, there is an implication shown that a Business Unit Representative or Liaison role will be needed as part of the solution.

Management Model Example #1

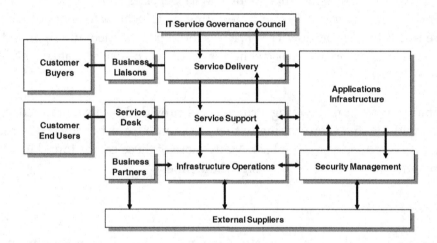

Management Model Example #2

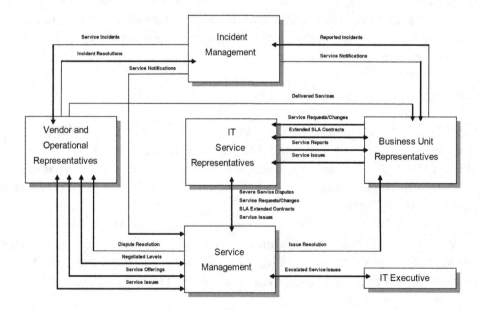

Technology Approach

This describes a high level view that shows what key tooling architectural components will needed to support the vision and how they interact. This is shown through a generic logical architecture model that describes the types of tools and data repositories that will be needed to support the Vision.

At this stage you will only identify the types of tools and data that will be part of the vision, not specific tool products, and databases. For assistance with this, refer to the Tooling Considerations chapter in this book for a description of ITSM tool types and requirements.

Governance Approach

This describes a high level view of the strategy that will be used to control and manage the future state Vision. See the chapter on Service Management Governance for more information.

Service Demand Factors

This describes a high level view of known business and IT events that will occur in the future that might have impact on the Vision, the Vision Approach and how the future state may operate. Think of these as major events that need to be considered. Some of them may actually be reasons for justifying the implementation effort in the first place. Examples of these might include:

- An impending acquisition or merger with another business or IT organization

- New business services that will be rolling out to customers

- Major increases in business volumes such as transactions, sales, or customer growth

- Service problems that are impacting profits or market share

- Impending regulatory requirement

- Other major business event

A key source to tap into for this will be the IT Service Management Focus Survey done earlier. This can provide much insight as to what may be happening with the business that is driving management interest in ITSM. Identification of the service demand factors can greatly support the reasons for an ITSM solution. It doesn't hurt to remind upper management of what is taking place within the business that drives the need for ITSM practices and management approaches.

Impact

As part of your strategy, you will need to help identify the impact the ITSM Vision will impose on the overall business organization. Here, you will be specifying what is changing as a result of implementing the vision. The table below provides an example of how you might indicate some of the impacts on your organization:

Area	Potential Impacts
Customer	Added value that will occur Changes in customer satisfaction Adding new customers Reporting service to customers
Organization	Must organize to deliver good service Decision making will be by service not technology
People	Changes in how people are rewarded Formalizing job roles and skills Customer driven mindset Need to formalize service communications Training and skill sets
Process	Changes in how role of processes are viewed and operated Changes in how processes are monitored and reported on Consolidation of similar processes Single process focus across stovepipe organizational pillars
Technology	Changes in how tools are selected Changes in how tools are integrated Architectural Changes
Governance	Collection and reporting on services Establishment of service targets Reporting on service quality
Culture	Service and Customer Orientation Understanding the Customer Prioritizing based on Customer Needs

Note that in the above example, the impacts described are very broad and general. At this stage, you are looking for the types of impacts that will need to be considered. You are not designing and answering how those impacts will be addressed. The goal is to understand the potential scope and breadth of what implementation of the Vision will mean to the organization.

Benefits

Much of the groundwork for this part of the effort has already been done for this with the Customer Needs, ITSM Focus and ITSM benefits surveys that were performed earlier. For this Work Task, you will simply be taking the information collected in those surveys, scoring and summarizing the benefits based on those results.

An inventory of benefits should be created that lists each benefit. When pulling this together, consider that benefits will fall into one of two categories:

Hard Benefits

This covers benefits that have direct and tangible value to the company. Examples of these include items such as:

- Reducing Cost
- Increasing Customer Satisfaction Ratings
- Lowering Headcount
- Reducing major service outages
- Delaying headcount or other resource acquisitions
- Meeting regulatory requirements
- Providing capability to support a new initiative

Soft Benefits

This covers benefits that have indirect value to the company. Examples of these include items such as:

- Reducing labor hours spent on non-value tasks
- Increasing overall service quality
- Increasing morale among IT staff
- More responsiveness to changes in business needs
- Achieving an industry quality award or milestone
- Understanding IT spend
- Better relationship with IT vendors and suppliers

Success Metrics

It is very important to make sure that you identify and agree a series of metrics that will outline the progress on whether the Vision itself has been achieved. This is a step that is also very critical to complete. These metrics will be used to monitor the success of the effort and to demonstrate what is being achieved.

The metrics serve a number of purposes. They focus the ITSM Implementation Team on aspects of the program that is very important. They also allow the team to communicate what is happening in terms of progress with Senior Management personnel. They can also serve as markers to trumpet the success of the program itself.

When Senior Management personnel come calling 6 months down the road to ask what has been achieved, you have a ready source and answer. Without this, the team may find itself constantly scrambling to prove their value over and over with upper management over time. Within the Vision Work Stage, everyone can agree on what the targets are for success early on and avoid questions and doubts later on.

When drafting the success metrics, there are several rules you may wish to consider now to avoid trouble later on. These are:

- Pick metrics that are measurable. There is nothing worse than getting everyone excited about a success metric and then finding out that it cannot be measured.

- Document assumptions and calculations as to how the metrics will be derived. This will avoid doubts and questions later on. Make sure there is agreement on calculations and assumptions as well as the metric itself.

- Do not pick metrics that will require lots of administrative overhead to collect. Make sure that results can be reported on without spending vast amounts of processing, data collection and summarization effort.

- Make sure you choose metrics that are not vague in nature. For example: something like "Percent morale improvement" may be pretty hard to measure versus "Employee satisfaction rating > 9.0". If you do choose a fairly vague metric, ensure that the supporting assumption and calculation explanation is rock solid.

- Do not choose metric targets that are never achievable. This becomes extremely discouraging to all team members.

The approach for drafting your success metrics should be something like the following:

1. Inventory the benefits as identified earlier.
2. Identify success factors for each benefit
3. Indicate one or metrics that will apply to each success factor
4. Identify a target value for each metric
5. Document how each metric will be calculated
6. Document any assumptions that apply to each metric

The table on the following page is an example of one way to indicate success metrics for your vision. In looking at this, you will notice that an additional column has been added for the target baseline marked as TBD.

The baseline refers to the current state of that metric value as it exists today. The target indicates the future desired state of that metric. The difference between the baseline and target outlines the improvement that is expected as a result of the ITSM Implementation Program. In this way, you will get a valuable before and after picture that can be used to demonstrate program success and keep the momentum going.

The task of populating the baseline column will occur in the next ITSM Assessment Work Stage. For this Work Stage you should leave it as TBD.

ITSM Program Success Metrics Example

Benefit	Success Factor	Metric	Calculation	Baseline	Target	Assumption
Increase Customer Satisfaction	Better satisfaction ratings	Monthly Customer Survey Score	Average survey score as reported monthly in the Customer Survey	6.2	> 9.0	None.
Reduce Costs	Reduce overall TCO for delivery and operation of IT services	IT Cost per employee per month	Total IT expenditures per month divided by number of company employees	$460	< $300	Will use expenditures as reported on monthly IT budget report. Will use employee count as reported in monthly HR statistics report.
Improve IT Service Quality	Reduce major service outages	Severity 1 level outages per month	As reported from corporate incident management tool	31	< 2	None.
Improve IT Service Quality	Reduce number of repeat incidents	Percent of incidents marked as REPEATED	Total REPEAT incidents as reported from corporate incident management tool divided by total number of incidents reported by that tool	42.5%	< 10% of total incident count	None.
Improve IT service delivery and support capabilities	Demonstrated adherence to best practices	ITSM Process Maturity Rating	Rating as assessed by 3rd party independent firm for ITSM Process Maturity	TBD	> 2.0	None.

Benefit	Success Factor	Metric	Calculation	Baseline	Target	Assumption
Better relationship with IT vendors and suppliers	Reduced complaints and contract issues	Number of vendor contract issues by each vendor	Sum of number of vendor contract issues rated as high and medium for each vendor	TBD	< 2 high rated issues; < 5 medium rated issues	Will create inventory of contract issues for each key vendor, prioritize these and review with each vendor.

Building the Communications Strategy Task

Developing and agreeing a Vision and Vision Strategy is actually the easier of the efforts in this Work Stage. The harder part is making sure that the rest of the business organization is aware of that vision and will agree that it is worthwhile. The purpose of developing a Vision Communications Plan is to make sure that you are identifying how you will communicate and agreeing your vision across the organization. By the end of this Work Stage, key decision makers within the business organization should agree that:

- There is a business problem that cannot wait to be solved

- The right people are involved in addressing this problem and leading the charge

- The Vision and Strategy to solve the problem are the correct one

If you miss any of the above, there will be a high risk that any further efforts to implement an ITSM solution may meet high levels of opposition or simply stall later down the road.

The key steps to getting the Strategy developed are:

- Identifying the impacts of the Vision and Vision Strategy being developed

- Identifying and engaging Stakeholders in the solution

- Identifying the benefits to be achieved by the Vision

- Developing key messages around the Vision and the Vision Strategy

- Developing an overall communications plan for the Vision and the Vision Strategy

The effort to accomplish the above tasks should is not to be underestimated. Experience has shown that communication work typically will take over 70% of Senior Executive time spent on this effort. It will most likely take an additional 20% of time of the Vision Team itself. It is highly recommended that a full-time Core Team member lead the Organization Track for this effort.

Refer to the chapter on Adapting to an ITSM Culture for more details on developing a Communications Plan and leading your organization through behavioral and cultural changes for the vision.

Identify Program Risks Task

Many of the potential Program risks can be drawn out of the things gathered and collected earlier. Items such as the ITSM Cultural Survey and Stakeholder Analysis work can yield potential risks and concerns. Another approach is to hold a working session and have participants jot down and discuss risks that they see.

Risks can fall into categories. Some of these include:

- Technology

- Culture

- Organization

- People and Skills

- Financial

- Transition

- Governance and Reporting

Examine each area of the Program as it is envisioned and determine whether potential risks may exist in them. A more diligent risk analysis will occur in the later ITSM Planning Work Stage. For this stage, the risks you are looking for are the big hitters. This represents those items that might be showstoppers or cause major delays in the implementation effort if not adequately addressed.

Simply inventory each risk found. Once listed, be sure to include a general mitigation action that will occur in the Program to mitigate it. For example, if one of the risks highlighted was "not enough ITSM skills", the associated mitigation action might indicate "Institute ITSM foundation level training program to all implementation staff prior to starting implementation work" as a potential mitigation action.

A sample template for describing risks and mitigation actions, using the example below, might look like the following:

Category	Risk	Impact(s)	Priority	Mitigation
Skills	Lack of ITSM skills	Program delays Poor implementation quality Potential for rework Higher implementation costs Lack of change acceptance	High	Institute ITSM Foundation training for all Program Participants prior to implementation start
Next Category	Next risk . . .			

Note that in the above example, both an impact and priority have also been added. This would be recommended as it highlights why the risk may be serious and the priority that with which it should be addressed.

In this Work Task, you will be building a table similar to the above as a way to identify the initial risks that were found. Make sure that the Governance Work Track eventually has a chance to review this, as it will be needed for estimating costs.

Estimating Program Costs Task

The purpose of this Work Task is to draft a set of cost estimates for the program itself. You might be asking at this point how this could be accomplished. An assessment hasn't even been done yet alone all the planning for what it will take to accomplish this!

For the Vision Work Stage, you are trying to identify large ballpark level cost items to begin to get a potential handle around what the Vision may cost to implement. This is for early estimating purposes only. It is used to justify the decision to proceed to the ITSM Assessment Work Stage, not the entire effort.

Towards that end, you will be building a high level cost model. A template for one is shown below. In this model, you will be listing a set of cost factors along with associated cost estimates and assumptions. This is not a detailed cost model for the entire program. To reinforce that concept with Senior Management, you will be plugging in estimated cost ranges versus single figures.

Assessment Work Stage Costs (Example):

Cost Factor	Estimate	Assumptions
Training	$x—$x	Assume n people at $x per head
Personnel Labor	$x—$x	Assume average hourly rate of $x times 170 (one month's labor hours) times n employees times number of months
Facilities Costs	$x—$x	Assume $x per employee for desk space and meeting rooms
Infrastructure	$x—$x	Covers PCs, tools and networks used to operate the program
Consulting	$x—$x	Assume general run rate of $x per month
Administrative	$x—$x	Ballpark secretarial and administrative assumptions here
Other	$x—$x	Specify your assumptions here . . .
TOTAL:	**$x—$x**	

Program One Time Costs (Example):

Cost Factor	Estimate	Assumptions
Training	$x—$x	Assume n people at $x per head
Personnel Labor	$x—$x	Assume average hourly rate of $x times 170 (one month's labor hours) times n employees times number of months
Facilities Costs	$x—$x	Assume $x per employee for desk space and meeting rooms
Infrastructure	$x—$x	Covers PCs, tools and networks used to operate the program
Hardware/ Software	$x—$x	Using typical average for various tool types or published industry estimates
Consulting	$x—$x	Assume general run rate of $x per month
Administrative	$x—$x	Ballpark secretarial and administrative assumptions here
Other	$x—$x	Specify your assumptions here . . .
TOTAL:	**$x—$x**	

Note that the above model has broken out the Assessment Work Stage costs separately. The reason for this is that you are seeking Senior Management go-ahead agreement to do the next Assessment Work Stage as a key outcome of this Work Stage. The funding commitment from Senior Management will only be for that Work Stage at this point in time.

Also make sure you review the outcome of the Risk Analysis Work Tasks done earlier. For these, you may wish to scan for items that look like they might be large cost hitters. For example, if there appear to be lots of cultural barriers to overcome, you may want to double some of your labor estimates.

During the ITSM Assessment and ITSM Planning Work Stages, you will be refining these cost estimates much more. It will be at the end of the Planning Work Stage that you will be asking Senior Management for the final go-ahead to actually start the ITSM Implementation Program and therefore need to be much more accurate in your estimates.

You may be asking what the point of running through this cost exercise is if we are mostly producing estimates that will be refined later. The purpose is to begin to shape and mold the potential costs for the Vision for Senior Management. If they have no appetite to spend anything on ITSM, let's find out about it now rather than waste more time later.

When building your estimates, take your best shot and then round solidly up adding what you feel is a reasonable contingency. Do not be overly concerned if you aim a bit high. One of two things will happen later. If costs come in lower than planned you can proudly announce to Senior Management that you found a cheaper way of accomplishing your goals. Otherwise you will have built a contingency for things that you might have missed or that reared themselves later in the ITSM Planning process.

Sources for estimates can also include industry papers or simply finding other companies that have been working with ITSM and sharing cost information. Local interest groups such as the IT Service Management Forum can be excellent places for this. Industry research groups such as Gartner and Forrester are also good sources to get hardware/software solution estimates.

Also note that this Work Task also includes a Work Step to draft the ITSM Assessment Work Stage Project Plan. For this, you can steal from the plan presented in this book in the ITSM Assessment chapter. If you will be using outside 3rd party consultants to help with that effort, now would be a great time to get them engaged and help you plan and cost the effort before they begin to run the consulting meter. By building the detailed plan for this now, you will be able to better identify the needed funding for the next work stage in your cost model.

Assemble the Vision Document Task

In this Work Task, the Vision Document itself is assembled and drafted. The format for the document can be driven off the Work Product Description shown earlier for this document. A recommended set of sections for it might include:

- Management Summary

- Vision Statement

- Key Benefits

- Vision Strategy and Key Timeframes

- Program Success Factors

- Program Risks

- Estimated Costs

Along with the Vision Document, you should also assemble a Vision Presentation. This should be relatively brief and extracted from the content in the Vision document. The presentation will be what is eventually shown to Senior Management and others to get the necessary buy-in and approval.

Obtain Vision Agreement Task

This Work Task culminates in the Senior Management buy-in and go-ahead authorization. At this stage, the go-ahead authorization is to continue with the ITSM Assessment Work Stage only. There will be additional meetings with Senior Management at the end of the ITSM Assessment Work Stage and ITSM Planning Work Stage before the final go-ahead is given to start.

The Work Steps within this Work Task will primarily focus on execution of the Vision Communications Strategy first leading up to the Senior Management meeting. This means executing those parts of the strategy deemed critical towards obtaining the buy-in to the next Work Stage. This will most likely involve shopping the Vision Presentation to different stakeholders to make sure the appropriate decision makers and influencers are on board with the vision.

It is fairly typical that you may have multiple versions of your Vision Presentation to present to different audiences. These variations are usually due to the levels of detail that different audiences might be looking for.

Customer Needs Survey

Introduction	This survey identifies customer needs and desires from their IT support and delivery services.
Purpose	Prioritizes and identifies what kinds of needs and requirements ITSM customers are most interested in.
Audience	This survey should be delivered to receivers of IT services.
Instructions	Rank each Need as: 0—Not important to me at all 1—Very minor importance to me 2—Somewhat important to me 3—Important to me 4—Very important to me 5—Absolutely critical to me Weight each Benefit as: 9 = Feel strongly about this 3 = Feel important to consider but not overly strong about this 1 = Use this if not any of the above Add comments as necessary to support interest ranking and weighting
Scoring	Sum rank scores for each benefit area Sum weight scores for each area Multiply rank by weight to get total score for each area Tally comments for each benefit area
Meaning	Sum of weight scores indicates emotional interests in particular benefit areas Sum of rank scores indicates general interest in each benefit area Sum of weighted scores indicates overall priority for each benefit area Comments can be used for input to further assessment efforts

Customer Needs Survey

Introduction   The following table should be filled out by those who receive IT services. Make sure to base your answers on what you want and value—not an assessment of how good or bad these things are today.

My Needs	Rank	Weight	Comments
I want IT to fully understand my requirements for what they should be providing to me			
I want to understand what IT can or cannot provide			
I want solutions delivered to me in a timely manner			
I want services to be available when I need them			
I want the capability to recover information in event of erasure or failure			
I want speed and responsiveness of services when I use them (i.e. minimal delays when enter key is hit)			
I want problems and incidents addressed in a timely manner when things go wrong			
I want IT support staff to have the appropriate level of skills to support me			
I want consistent processes for procuring IT solutions and services			
I want protection from security threats			
I want services are quickly recovered when outages do occur			
I want services to be available to me in the event of a major business outage or disaster			
I want timely notification if a service disruption is about to occur			
I want a clear explanation of why an outage occurred without long winded answers or cryptic technical jargon			

My Needs	Rank	Weight	Comments
I want timely notification about IT activities and plans that might impact me			
I want estimates of costs when new services or service changes are proposed			
I want courteous IT support staff			
I want to make sure that easy-to-make errors on my part do not result in a major outage or service disruption			
I want a clear understanding of my IT costs			
I want my problems and requests to be always followed up on without having to make repeat calls to check up on things			
I want to receive timely notifications about progress on longer term problems and requests			
I want to get my work done without stumbling over technology issues			
I want confidence that the services delivered to me are compliant with corporate or industry regulations			
I want periodic information about how well services were historically delivered to me			
I want to have enough processing resources to get my work done effectively (i.e. PCs, Printers, etc.)			
I want to get and receive IT services with a minimum amount of bureaucracy			
Other wants—please list			

IT Service Management Focus Survey

Introduction	This survey identifies focus areas that may be driving the IT interest towards implementing ITSM solutions.
Purpose	Prioritizes and identifies where key interests in ITSM are coming from among various IT stakeholders.
Audience	This survey should be given to IT staff and management.
Instructions	Rank each Focus Area as: 0—No interest 1—Should be noted but no strong interest 2—Somewhat interested in this 3—Interested in this 4—Strong interest in this 5—Very strong interest in this Weight each Focus Area as: 9 = Feel strongly about this 3 = Feel important to consider but not overly strong about this 1 = Use this if not any of the above Add comments as necessary to support interest ranking and weighting
Scoring	Sum rank scores for each focus area Sum weight scores for each area Multiply rank by weight to get total score for each area Tally comments for each focus area
Meaning	Sum of weight scores indicates emotional interests in particular focus areas Sum of rank scores indicates general interest in each focus area Sum of weighted scores indicates overall priority for each focus area Comments can be used for input to further assessment efforts

IT Service Management Focus Survey

Introduction

The following table should be filled out by survey participants. Please add comments as to why you selected the focus area rank and weight.

Focus Area	Examples	Rank	Weight	Comments
Regulatory	Sarbanes-Oxley IT Audits			
Financial	Cost reduction Charging for services Setting IT Budgets Recognizing where IT costs are used Cost modeling			
Service Reliability	Fixing problems and incidents Stopping/reducing service outages			
Efficiency	Restoring service in a timely manner Handling operational tasks faster Changing labor balance to more hours spent on value-added tasks versus non-value tasks Reduce firefighting and internal finger-pointing Faster response for support of new business and IT initiatives			
Customer Satisfaction	Reducing/stopping customer phone calls and complaints Gaining customers and market-share Delighting customers			

Focus Area	Examples	Rank	Weight	Comments
Quality Initiative	Striving for quality award such as Malcolm Baldridge or other certification Corporate quality initiative programs such as Hoshin or Six-Sigma Internal desire to strive towards best practices			
Support For Impending New Systems Or Services	Building an infrastructure to support a major new application system or service Increasing levels of support services to meet new business demands			
Major Business Change	Pending merger or acquisition that must be supported Consolidation of IT services or business units Opening of a new processing center			
Outsourcing Considerations	Better management of outsourcing vendors or out-tasked areas Integrating services from multiple vendor providers Identifying requirements for potential outsourcing vendors Developing service plans for bringing IT services back in-house Planning for off-shoring services			
Other	Please describe what this is.			

Service Management Benefits Survey

Introduction	This survey identifies benefits areas that stakeholders would like to see from ITSM solutions.

Purpose	Prioritizes and identifies what kinds of benefits from ITSM stakeholders are most interested in.

Audience	This survey should be given to all stakeholder staff and management as well as company Senior Management.

Instructions	Rank each Benefit as: 0—No interest 1—Should be noted but no strong interest 2—Somewhat interested in this 3—Interested in this 4—Strong interest in this 5—Very strong interest in this Weight each Benefit as: 9 = Feel strongly about this 3 = Feel important to consider but not overly strong about this 1 = Use this if not any of the above Add comments as necessary to support interest ranking and weighting

Scoring	Sum rank scores for each benefit area Sum weight scores for each area Multiply rank by weight to get total score for each area Tally comments for each benefit area

Meaning	Sum of weight scores indicates emotional interests in particular benefit areas Sum of rank scores indicates general interest in each benefit area Sum of weighted scores indicates overall priority for each benefit area Comments can be used for input to further assessment efforts

IT Service Management Benefits Survey

Introduction   The following table should be filled out by survey participants.

CIRCLE ONE OF THE FOLLOWING:

My Company Role Is: STAFF   MIDDLE MANAGEMENT   SENIOR MANAGEMENT

CIRCLE ONE OF THE FOLLOWING:

My Association Is: I PROVIDE IT SERVICES   I SUPPORT IT SERVICES   I RECEIVE IT SERVICES

Benefits Area	Benefit	Rank	Weight	Comments
Financial	Cost justified IT infrastructure and services			
Financial	Reduction in IT costs			
Financial	Ability to charge fairly for IT services			
Financial	Protected investments in new IT solutions			
Employee	Common taxonomy and service culture among IT staff			
Employee	Reduction in labor spent on non-value tasks			
Employee	Motivated staff			
Employee	Increased operational expertise and skills among staff			
Employee	Gained visibility and reputation for work done			
Employee	Obtain clear understanding of roles and responsibilities			
Employee	Increased level of operational knowledge with best practices approaches			
Employee	Certification in IT Service Management			

Benefits Area	Benefit	Rank	Weight	Comments
Innovation	Ensure new IT solutions can be deployed and operated on a daily basis			
Innovation	Quickly position to support new IT solutions and technologies			
Innovation	Develop operational solutions that provide unique value to other business areas			
Innovation	Support selection of appropriate operational support technologies			
Business	Ensure operational management activities and priorities are aligned with business activities and priorities			
Business	Ensure operational support and services adapts quickly to changing business needs			
Business	Reduce frequency of service outages			
Business	Reduce duration of service outages			
Business	Improve relationships between IT service providers and their customers			
Business	Improve customer satisfaction			
Business	Ensure business services can operate in the event of a major IT outage			
Business	Reduce or eliminate time spent by business unit staff addressing or dealing with IT problems and issues			
Business	Foster understanding as to what IT services can or cannot do			
Business	Demonstrate high level of IT service support quality to gain customer market share			
Business	Demonstrate compliance with external audit or regulatory agencies			

Benefits Area	Benefit	Rank	Weight	Comments
Business	Holding IT accountable for the quality of services they deliver			
Business	Development of effective service support and delivery models for off-shore, or outsourcing			
Business	Provide IT services based on business terms versus technological silos			
Internal	Improve metrics and management reporting			
Internal	Reduce or eliminate complaint calls from customers or company management staff to IT personnel			
Internal	Raise visibility of IT within the business organization			
Internal	Position IT to operate as a shared services provider or self-funding business unit			
Internal	Eliminate finger pointing and confusion over roles and responsibilities			
Internal	Get out of firefighting mode			
Internal	Increase maturity of processes			
Internal	Ensure IT services can be recovered in the event of a disaster			
Internal	Implement clear understanding of operational services and service targets among IT units and staff			
Internal	Obtain more efficient IT management and support processes			
Internal	Institute program for continual operational improvement			
Internal	Obtained clear view of current IT capabilities and services			
Internal	Obtaining a recognized quality award (i.e. Malcolm Baldridge)			

Benefits Area	Benefit	Rank	Weight	Comments
Internal	Ensuring appropriate balance of capacity between expenditures and needs			
Internal	Obtaining awareness of upcoming business changes or events that will have impacts on ability to deliver services			
Internal	Positioning for a major business application, merger, acquisition or consolidation			
Internal	Effective management of vendor service providers and outsourcers			
Internal	Improved ability to assess impacts of changes			
Internal	Better manage increasing complexity in technologies and services			
Other	Please describe here			

IT Service Management Cultural Survey

Introduction

This survey identifies how people view the current IT service delivery culture as it exists today.

Purpose

Identifies where cultural issues may exist so that efforts to overcome these can be recognized and incorporated into the ITSM implementation effort.

Audience

This survey should be given to all stakeholder staff and management as well as company Senior Management.

Instructions

Circle the number that reflects how close you feel the current organizational behavior matches the statement to the left or the right.

Scoring

Obtain the average for each table line item. Round to nearest whole value. Sum all the values for each line item.

May also wish to compare scores between users and providers as well as management and staff.

You may also wish to view scores by cultural issue area. The question ranges fall into the following categories. You can average scores for each category to get locate the type of cultural issue areas that may cause more problems than others. The categories are:

Question Row	Category
1-3	Product and Service Planning
4-7	Measuring Performance
8-12	Attitude Towards Customers
13	Quality of Products and Services
14-15	Marketing Focus
16-17	Process Management Philosophy
18	Product Service and Delivery Philosophy
19-20	Customer Focus
21-26	Operational Culture

Meaning   Total score meanings apply per the table below:

Score Range	Meaning
0-26	Indicates major cultural issues that will need to be overcome High risk for a successful ITSM implementation effort with little chance for success
27-52	Indicates many issues that need to be overcome Plan on significant amounts of ITSM effort spent on organizational change activities High risk for a successful ITSM implementation effort with low or medium chance for success
53-78	Indicates many issues still exist Should examine where scores were highest to pinpoint areas where problems might lie Medium high risk with uneven chance for success
80-104	Some issues exist Should examine where scores were highest to pinpoint areas where problems might lie Moderate risk for success
105-130	Few issues exist Few risks that cultural issues will stand in the way of a successful ITSM Implementation effort

IT Service Management Cultural Survey

Introduction   Circle the number value in each row that is closest to your view of the current organization.

CIRCLE ONE OF THE FOLLOWING:

My Company Role Is: STAFF   MIDDLE MANAGEMENT   SENIOR MANAGEMENT

CIRCLE ONE OF THE FOLLOWING:

My Association Is: I PROVIDE IT SERVICES   I SUPPORT IT SERVICES   I RECEIVE IT SERVICES

Reference	Most Like Us	Score	Most Like Us
1	Short term focus	1 2 3 4 5	Long term focus
2	Reactionary management	1 2 3 4 5	Prevention based management
3	Management by objectives planning process	1 2 3 4 5	Customer driven strategic planning process
4	Bottom line financial results	1 2 3 4 5	Customer satisfaction
5	Quick return on investments	1 2 3 4 5	Strategically focused to gain Market share
6	Short term profitability	1 2 3 4 5	Long term profitability
7	Technical orientation	1 2 3 4 5	Customer orientation
8	Productivity by IT organization unit	1 2 3 4 5	Productivity across all IT organization units
9	Customers are irrational and a pain	1 2 3 4 5	Voice of the Customer is very important to us

Reference	Most Like Us	Score					Most Like Us
		1	2	3	4	5	
10	Customers are a bottleneck to profitability	1	2	3	4	5	Professional treatment of customers is required
11	Hostile and careless towards customers	1	2	3	4	5	Courteous and responsive to our customers
12	Take it or leave it attitude towards customers	1	2	3	4	5	Empathy and respectful attitude towards customers
13	Products and services provided based on current technologies and organizations	1	2	3	4	5	Products and services provided according to customer requirements and needs
14	IT acts as if in a sellers' market	1	2	3	4	5	IT acts as if in a competitive market
15	Very weak focus on customer satisfaction	1	2	3	4	5	Very strong focus on customer satisfaction
16	Focus on error and defect detection	1	2	3	4	5	Focus on error and defect prevention
17	Many different forms of similar processes in different organizations	1	2	3	4	5	Processes span across organizations
18	It is okay for customers to wait for products and services	1	2	3	4	5	It is best to provide fast time to market products and services
19	People are the source of problems and are burdens on the organization	1	2	3	4	5	People are our greatest resource
20	We are product driven	1	2	3	4	5	We are customer driven
21	We make management decisions based on opinion	1	2	3	4	5	We make management decisions based on data
22	We operate by fighting the next management crisis	1	2	3	4	5	We operate with continuous process improvement
23	We manage by fear and intimidation	1	2	3	4	5	Staff is empowered to make the best decisions
24	Customers, suppliers and process owners have nothing in common	1	2	3	4	5	Strong teamwork between suppliers, process owners and customers

Reference	Most Like Us	Score	Most Like Us
25	People are career driven and work independently	1 2 3 4 5	People work in teams to accomplish objectives
26	Rewards are given to those who demonstrate the highest technical competence	1 2 3 4 5	Rewards are given to those who serve their customers best and maintain high quality service

IT Service Management Labor Survey

Introduction	This survey identifies how IT labor is currently spending time on value and non-value tasks.
Purpose	Identify the current level of effort spent on non-value work.
Audience	This survey should be given to all IT service delivery and service support staff and management.
Instructions	Indicate the approximate percent of your time spent on the following activities per week. Make sure the sum of your answers = 100%.
Scoring	Obtain the average percentage value for each line item. Sum the percentage values by each labor category. These are as follows:

Labor Category	Activity Type
Value-Add Labor	Developing Solutions Representing Solutions Researching
Non-Value Labor	Monitoring Supporting
Overhead Labor	Administering Maintaining Managing

Meaning	The goal is to maximize Value-Add Labor to greatest extent possible and reduce Non-Value Labor to greatest extent possible. Some warning signs might be a Non-Value Labor score higher than 30% and/or an Overhead Labor score higher than 15%. High non value or overhead indicates areas that may benefit from implementation of an ITSM solution.

IT Service Management Labor Survey

Introduction — Indicate the approximate percent of your time spent on the following activities per week. Make sure the sum of your answers = 100%.

Activity	Examples	Approximate Percent
Monitoring Solutions	Watching terminals, reviewing system logs for issues, checking job status, performing operational tasks such as job initiation or server startup/shutdown	
Developing Solutions	Designing new IT solutions, developing scripts and code, migrating code to production libraries, database or network design, developing automated operational scripts, working on an IT project, implementing a major software upgrade, developing capacity, availability and continuity plans	
Administrating	Collecting data for IT reports, filling out timesheets, producing status reports, producing budget and expense reports, billing for IT services, purchasing IT resources and supplies, developing IT budgets	
Maintaining Solutions	Applying hardware or software maintenance, applying software patches and releases, working on processing facilities improvements	
Managing	Overseeing and reviewing the work of others, conducting employee review meetings, recruiting additional staff, managing a project effort	
Researching	Performing hardware/software selections, meeting with vendors and architects, conducting industry surveys, estimating costs for a solution	
Supporting Solutions	Responding to incidents, resolving service problems, responding to customer calls, responding to management calls for status on a problem or issue	
Representing Solutions	Identifying new customers and IT services, identifying customer needs and service levels, consulting with end users and customers to identify needs and potential solutions, assisting procurement with IT contracts and vendor issues	
	TOTAL:	100%

IT Service Management Alignment Tool

This tool is derived using some of the Six-Sigma Quality Function Deployment (QFD) techniques to prioritize improvement efforts based around customer demands. In the case of ITSM, IT personnel may try to "guess" what ITSM processes should receive the most attention. The purpose of this tool is to help guide those decisions more in line with what customers are really asking for.

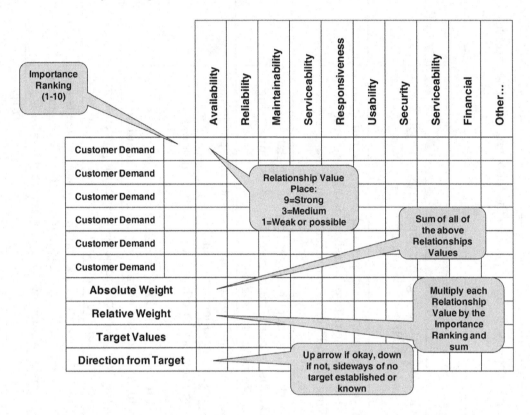

The tool is essentially a series of corresponding matrices that are completed and filled out by the Vision Team. The following chart illustrates how this tool is put together:

The above can be assembled on a spreadsheet. The Customer Demand elements will come from the Customer Needs Survey done earlier. The column headings at the top represent common IT service characteristics. An OTHER column has also been included if you wish to add other characteristics not included above. The top column headings can be referred to as Critical to Quality items or CTQs.

Essentially, you will first be identifying which CTQs relate most for each customer demand, weighting and ranking which CTQs have the most weight and then using the scoring criteria to identify which ITSM areas may need the attention first.

By linking things in this manner, you will have demonstrated how user needs were translated into ITSM priorities. With this, you avoid the guesswork in selecting which processes to focus on most. You can also show non-IT personnel how your decisions were arrived at in a diligent manner.

The Work Steps for using this tool are as follows:

Step	Action
1	Fill in the Customer Demands rows. These can be taken from the Customer Needs Survey done earlier.
2	Fill in the importance ranking. This is calculated for each demand item by taking the average of all rankings for it from the Customer Needs Survey and then doubling that value. Round to the nearest value to get a 1-10 score as needed.
3	Confirm the CTQs that will be used (i.e. Availability, Reliability, etc.). Add any additional ones if you see fit.
4	Identify target values for each CTQ if known. These just need to be general, not specific to an application or service. For example, you might list 99.8% as a general Availability target based on past experience or possibly something that was listed as a Customer Demand. If no known metric exists, then just put N/A in Target Column.
5	Identify a relationship score for each column/row that established the strength of the relationship between the CTQ and the Customer Demand. The only possible values are 9 if there is a strong relationship, a 3 if there is a medium relationship or else place a 1 there. Since this exercise should be done as a team, determine how there will be consensus on this value such as using a simple vote.
6	Determine the absolute weight for each CTQ by summing the scores for all Customer Demands in the associated column.
7	Determine the relative weight for each CTQ by multiplying each relationship value by the associated importance ranking and then summing the results. For example, if the first Customer Demand/ Availability CTQ had a score of 9 and the importance ranking for it was 5, then the result would be 45. Proceed to the next customer demand and repeat, then sum all customer demands for the Availability column to get the relative weight value.
8	Place arrows in the direction from target row. An UP ARROW indicates an opinion as to whether the target is currently met; a DOWN ARROW indicates that the target is currently not met; a SIDEWAYS ARROW indicates that there is no knowledge of a target being met.

The table on the following page illustrates how the scores from this can be interpreted. Note that the PRIMARY column indicates a strong relationship between the CTQ and the listed ITSM processes.

The SECONDARY column indicates that there is a moderate relationship between the CTQ and the listed ITSM processes.

The processes listed are only recommended. There may be strong reasons to include other process areas as a higher priority based on:

- A particular Customer Demand Item listed

- Known service issues that exist within IT

- Service Demand Factors that indicate future events that must be prepared for

Scoring Results for the ITSM Alignment Tool

CTQ	Definition	Primary	Secondary
Availability	Ability of a component or service to perform its required function at a stated instant or over a stated period of time. The proportion of time that the service is actually available for use by the Customers within agreed service hours.	Availability Management Problem Management	Incident Management Release Management Change Management Capacity Management IT Service Continuity Management Security Management
Reliability	The ability for a service to continue over time even through the occurrence of one or more incidents.	Availability Management Problem Management	IT Service Continuity Management
Maintainability	Ability of company IT personnel to maintain services or service components so that they can continue to perform their stated functions.	Service Level Management Configuration Management	Change Management Release Management Availability Management
Serviceability	Ability of 3rd party vendor personnel to maintain services or service components so that they can continue to perform their stated functions.	Service Level Management Configuration Management	Change Management Release Management IT Service Continuity Management Financial Management
Responsiveness	Freedom from service delays or slowdowns that hinder the ability to get work done in a timely manner. Also includes responsiveness of IT to business changes and requests.	Service Level Management Capacity Management Incident Management	Change Management Release Management Service Desk
Usability	Ease of use for delivered IT services and equipment.	Service Level Management Release Management	Problem Management Security Management Service Desk

CTQ	Definition	Primary	Secondary
Security	Capabilities to provide confidentiality, integrity and access to business data, applications and services.	Service Level Management Security Management Incident Management	Problem Management Change Management Release Management Availability Management
Financial	Capabilities to fairly account for, budget and charge for delivery and support of IT services.	Service Level Management Financial Management Configuration Management	Availability Management Capacity Management IT Service Continuity Management
Other . . .	Place your description here	Identify primary areas here	Identify primary areas here

ITSM Visioning Workshop Strategy

Introduction   This describes an approach for quickly getting consensus around a vision statement for ITSM services in a workshop setting

Workshop   A proposed workshop is described that will allow a vision statement to be obtained in under 2 hours.

Agenda   Recommended agenda is as follows:

Agenda Item	Approximate Time
Introductions	15 minutes
Purpose and Overview	5 Minutes
Values Assessment	20 minutes
Vision Drivers	20 minutes
Brainstorm Vision	45 minutes
Summarize and Wrap-Up	10 Minutes

Introductions   Facilitator introduces self
Each member says name, role in company
Each member states personal vision thoughts in a simple sentence
Facilitator notates personal vision thoughts on a flip chart

Purpose and   Review purpose and outcome of session
Overview   Explain the session agenda and what will be done

Values   Review Corporate Vision Statement if one exists
Assessment   Each member gets 10 sticky dots
Values List is shown on the wall (See Attachment A)
Members modify and agree the Values List as needed
Members spend however many dots they want on any values in the list
Facilitator ranks these by which ones got the most dots
Prioritized list is reviewed with group

Vision Drivers	Vision Statement Definition is reviewed with group (See Attachment B) Driver questions are shown on the board or flipchart (See Attachment C) Facilitator reviews these one at a time with members Key points are documented underneath each appropriate question
Brainstorm Vision	Each member gets a yellow sticky pad Members get 15 minutes to write down words or sentences that they feel should be included in the vision statement Members place one word or sentence per each sticky paper Facilitator then reads these individually and makes sure concept is understood by group Facilitator then places these on a whiteboard and initially groups them into broad categories based on similarity Sticky items are then placed together in like statement structures Statements are then modified and streamlined based on group discussion Use dots techniques similar to before if group cannot agree among several sentence options
Summarize and Wrap Up	Review vision statement against results of value assessment to confirm agreement Review vision statement for alignment with corporate vision statement if one exists Note any issues or concerns if these have arisen Describe any next steps to be done

Attachment A—Values List

Overview

This describes the Values List used for the workshop from an IT perspective.

Place Dots Here	Value
	Reduce Risk
	Increase Revenue
	Decrease Costs
	Avoid Costs
	Improve IT Service Quality
	Improve Productivity Of IT Staff
	Decrease Time To Market
	Improve Customer Service
	Provide Competitive Advantage
	Become Recognized Market Leader
	Other (describe here . . .)

Attachment B—Vision Statement Definition

Overview	This describes the definition of a vision statement and provides some examples that could be used in the workshop.

Vision Statement Defined	Mutually agreed statement of "where we want to be" Based on looking at forward IT and business objectives Should describe the aim and purpose for IT Characteristics include: Clarifies direction of the Service management program Motivates people to take positive action Coordinates the actions of people, process, technology and governance Outlines the views of Senior Management Can be explained in less than 2 minutes to each stakeholder Not specific to products or a way of operating

Good Vision Statement Examples	Support COMPANY NAME business goals of transforming itself into a leading information company by providing the highest quality IT service infrastructure that delivers the most reliable, accurate and secure (BUSINESS SERVICE OR PRODUCT) services in the world. We will set the standard for how to operate a global IT infrastructure that is admired around the world—most importantly—by our customers as a reason to buy COMPANY NAME services. We will be the preferred supplier of IT services to the business.

Attachment C—Vision Statement Discussion Questions

Overview

The following questions may be used to help workshop members think about possible vision statements:

Who are our customers?

How should we appear to our customers?

How should we appear to our stakeholders?

What must we excel at with our customers and stakeholders?

How will we sustain our ability to change and improve?

Chapter 7

ITSM Assessment Work Stage

"An individual without information cannot take responsibility. An individual who is given information cannot help but take responsibility."
—Jan Carlzon, President, Scandinavian Airlines

The main goal of the ITSM Assessment Work Stage is to identify the gaps that exist in the organization between its current state and the future state as envisioned in the ITSM Vision Work Stage. The assessment efforts will be looking at processes, technology capabilities, organization readiness and governance control practices as they exist today within your organization.

The assessment itself will yield three main outcomes:

Assessment Results

These will identify what gaps exist between current and future state. They also provide a benchmark between current state and ITSM best practices.

Assessment Recommendation Actions

This is an inventory of recommended actions to be taken to overcome the gaps identified. These actions should be prioritized based on the outcomes of the Program Visioning Work Stage described earlier. The recommendations should also include impact statements that describe what might be expected if they are not implemented.

Recommended Initial Win Projects

This is an inventory of Recommended Initial Win Projects to be undertaken that appear to provide noticeable short term benefits to IT and the business organizations they support. An Initial Win Project is a sub-project within the ITSM Implementation Program. It provides real benefits and must be completed within a year. Initial Win Projects should indicate which Assessment Recommendation Actions they are addressing.

ITSM Assessment Work Stage Approach

If we look at the Assessment effort over an approximate 8-9 week timeline, the key Work Tasks might be timed to look like the following:

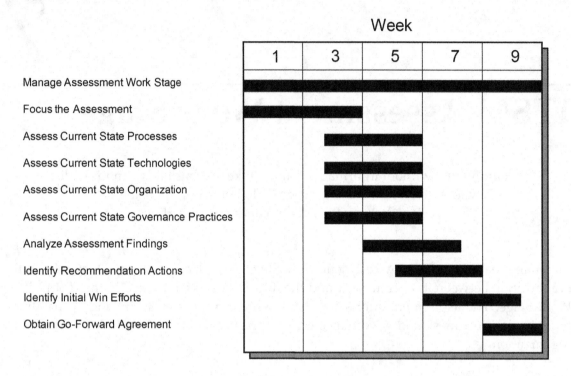

A detailed Work Breakdown Structure is depicted on the following page:

ITSM Assessment Work Stage—Work Breakdown Table

R=Responsible
A=Assists
C=Consults
I-Informed

Work Task	Work Step	Org	Proc	Tech	Gov	Mgt	Work Product
	Manage Assessment Work Stage activities	A	A	A	A	R	
Manage Assessment Work Stage	Report Assessment status	A	A	A	A	R	Status Reports
	Coordinate and schedule Assessment meetings	A	A	A	A	R	
	Integrate Assessment activities across all work tracks	A	A	A	A	R	
	Review ITSM Vision Work Stage results	A	R	A	A	A	
	Identify key areas of concern	A	R	A	A	A	
	Identify list of information items to collect	A	A	A	A	R	Information Collection Inventory
Focus The Assessment	Assemble assessment materials	A	A	A	A	R	
	Finalize Assessment Strategy	A	R	A	A	A	
	Identify additional stakeholders and initiatives	R	A	A	A	A	Updated stakeholder lists
	Schedule and conduct Assessment Stage Kickoff Session	R	A	A	A	A	

153

Work Task	Work Step	Org	Proc	Tech	Gov	Mgt	Work Product
Assess Current State Processes	Collect and review process related information items	C	R	C	C	I	Set of current state process related documentation
	Conduct Process Assessments	C	R	C	C	A	ITSM Process Assessment Results
	Inventory Process Findings	C	R	C	C	I	
Assess Current State Technologies	Collect and review technology related information items	C	C	R	C	I	Set of current state technology related documentation
	Conduct Technology Assessments	C	C	R	C	A	ITSM Technology Assessment Results
	Inventory Technology Findings	C	C	R	C	I	
Assess Current State Organization	Collect and review Organization related information items	R	C	C	C	I	Set of current state organization related documentation
	Conduct Organization Assessments	R	C	C	C	A	
	Conduct Stakeholder Analysis	R	C	C	C	I	Stakeholder Analysis
	Inventory Organization Findings	R	C	C	C	I	ITSM Organization Assessment Results
Assess Current State Governance Practices	Collect and review Governance information items	C	A	C	R	I	Set of current state governance related documentation
	Conduct Governance Assessments	C	A	C	R	A	
	Conduct Baseline Analysis	C	A	A	R	I	Baseline Program Success Factors
	Inventory Governance Findings	C	A	C	R	I	ITSM Governance Assessment Results

Work Task	Work Step	Org	Proc	Tech	Gov	Mgt	Work Product
Analyze Assessment Findings	Identify gaps between current practices and future state vision	A	R	A	A	I	
	Determine assessment maturity scores and other ratings	A	R	A	A	I	
	Inventory assessment findings	A	R	A	A	I	
	Review assessment findings with key stakeholders	R	A	A	A	I	
	Update findings based on stakeholder feedback	A	R	A	A	I	Assessment Findings Analysis
Identify Recommendation Actions	Identify and inventory recommendation actions	A	R	A	A	I	
	Cross reference recommendations with each ITSM process area	A	A	A	R	I	
	Identify benefit levels for each recommendation action	A	R	A	A	I	
	Identify effort levels for each recommendation action	A	R	A	A	I	
	Identify barrier levels for each recommendation action	R	A	A	A	I	
	Identify impacts of not implementing each recommendation	A	R	A	A	I	
	Score and prioritize recommendation actions	A	A	A	R	I	
	Review recommendation actions and scoring with key stakeholders	R	A	A	A	I	
	Update recommendation actions and scoring based on stakeholder feedback	A	R	A	A	I	Recommendation Inventory

Work Task	Work Step	Org	Proc	Tech	Gov	Mgt	Work Product
	Group recommendation efforts into Initial Win activities	A	R	A	A	I	
	Cross reference each Initial Win with each ITSM process area and set of recommendation actions	A	R	A	A	I	
	Document Initial Win descriptions	A	R	A	A	I	
	Estimate Initial Win timelines and resource efforts	A	R	A	A	I	
	Document completion criteria for each Initial Win	A	R	A	A	I	
	Cross reference each Initial Win to the recommendations that they will address	A	R	A	A	I	
	Review Initial Wins with key stakeholders	R	A	A	A	I	
Identify Initial Win Efforts	Update Initial Wins descriptions based on stakeholder feedback	A	R	A	A	I	Initial Wins Descriptions
	Update initial implementation plans and estimates	A	A	A	A	R	Updated plans and estimates
	Develop work plan for ITSM Planning Stage effort	A	A	A	A	R	
	Finalize Program Charter	A	A	A	A	R	ITSM Program Charter
	Draft management presentation	A	R	A	A	C	Management Presentation
Obtain Go-Forward Agreement	Update plans and assessments based on any management feedback	A	A	A	A	R	
	Obtain management approval to go forward with ITSM Planning Work Stage effort	A	C	C	C	R	Go-Forward Agreement

Assessment Key Work Products

The table below presents each Work Product produced by the ITSM Assessment Work Stage that is a milestone and provides a further description of it. Other Work Products are described further in the ITSM Assessment Work Task Considerations section of this chapter.

ITSM Assessment Key Work Products

Work Product	Description
ITSM Process Assessment Results	Highlights gaps between current ITSM practices and industry recognized best practices May include a Process Maturity score rating Process assessment findings
ITSM Organizational Assessment Results	Highlights organizational readiness for change Highlights skill gaps Identifies current ITSM roles and responsibilities (using an ARCI index) May include a cultural assessment Organizational assessment findings
ITSM Technology Assessment Results	Inventories current ITSM related tool sets and key data repositories mapped against ITSM functions Highlights tool and data gaps May include opinion on how well current tools/data might support ITSM activities Technology assessment findings
ITSM Governance Assessment Results	Highlights gaps between current governance practices and industry recognized best practices Governance assessment findings Inventory of reported metrics (if they exist)
Stakeholder Analysis	Identifies stakeholder groups in more detail Identifies needs by each stakeholder group Prioritizes stakeholder perceptions and needs
Baseline Program Success Factors	Current measurement results for the Program Success Factors developed in the ITSM Vision phase Comparison to industry benchmarks or other targets that match the vision Corresponding measurement plan that outlines how data for each factor is collected and tracked
Assessment Findings Analysis	Integrates and inventories findings from each assessment: Process, Technology, Organization and Governance Identifies each finding with a finding reference number Includes determined process maturity ratings and other relevant assessment scores

Work Product	Description
Recommendation Inventory	Integrates and inventories recommendations from each area: Process, Technology, Organization and Governance Lists benefits of implementing each recommendation in terms of high, medium or low Lists estimated effort levels for implementing each recommendation in terms of high, medium or low Lists cultural and organization barriers for each recommendation in terms of high, medium or low Identifies impacts of not implementing each recommendation Documents other assumptions and comments specific to each recommendation Identifies a priority score for each recommendation based on a combination of effort, benefits, barriers and impact
Initial Wins Descriptions	Describes each Initial Win effort Cross references each Initial Win with the ITSM process area that will own it Identifies estimates of work effort and timelines Describes completion criteria for each Initial Win Cross references each Initial Win to the recommendations that they will address
ITSM Program Charter	Consolidates findings from both Assessment and Vision Work Stages Provides a going forward recommendation as to ITSM implementation scope and overall strategy Presents a business case for implementing an ITSM program Updates estimated costs for coming implementation efforts Identifies expected results over time via the expected implementation effort Identifies key staffing needs and expectations for the implementation effort
Go-Forward Agreement	Documented management approval to proceed to the next phase of effort Approved budget for the next phase of effort Identified staffing team for next phase of effort Identified consultants if these will be used for the next phase of effort

ITSM Assessment Implementation Team Organization

The recommended ITSM Implementation Team Organization is pictured below:

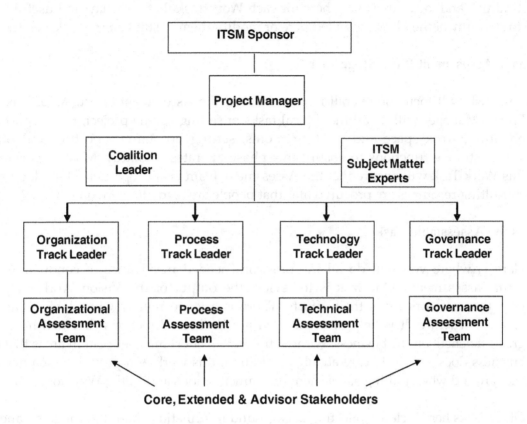

Drawing from the earlier chapter on Program Organization, the Assessment team organization will be a smaller subset. Note that the structure allows for the four Work Tracks: Organization, Process, Technology and Governance just as in the prior Work Stage. The teams that work within each track should consist of Core, Extended and Stakeholder types of individuals as described earlier.

While the selection of the team members is not as large or broad as the later implementation stages, a wider number of Extended and Advisor team players will be needed to provide assessment input and review.

For the Assessment efforts, it is recommended that an outside party be used. This is to ensure objectivity in the results. In addition, outside parties will typically bring in their own assessment criteria, approach and ITSM expertise. This avoids having to build these things as part of the assessment effort.

ITSM Assessment Work Task Considerations

This section describes each of the ITSM Assessment Work Tasks in a little more detail. Practical tips and ideas are shown here for each Work Task that you may find useful. Feel free to steal any of the ideas presented here or modify them to suit your specific needs.

Manage Assessment Work Stage Task

This task will focus on overall management of the Assessment effort. As before, the Project Manager will do all the typical tasks of setting up the project, time reporting, obtaining work space and meeting rooms, setting up status reporting and other administrative functions to ensure this phase operates smoothly. Most importantly, this Work Task will ensure that the Assessment Team is appropriately staffed, outside consulting resources are procured and that people are ready to work on it.

Focus the Assessment Task

Here is where you will do all the pre-work needed in advance of conducting the actual assessments. One task is to review the output of the Vision Work Stage to help the assessors focus their efforts. There is a lot to review when looking at the IT infrastructure. It is helpful to steer efforts in those directions that will have the greatest benefit on the business. Without this, much effort may be spent in areas that the business does not really care about. In addition, this work will provide much needed background when pulling assessment recommendations and Initial Wins together.

Other tasks here include collecting background information about the company and IT infrastructure for those doing the assessments. Examples of these would include:

- Organization Charts

- Existing Process Documentation

- Capacity and Performance Reports

- Service and Operational Level Targets

- IT Strategies and Plans

- Reporting Metrics

- Chargeback Reports

- IT Financial Budgets and Reports

- Problem and Incident Reports

- IT Continuity Plans

- IT Policy Documentation

- IT Service Reports

- Management Tool Architectures

- IT Governance Policies and Strategies

- Change Management Forms and Reports

- IT Issues Logs

- Service Improvement Plans

There may also be pre-assessment surveys to be done as requested by outside parties who will be doing the assessment. These will need to be distributed to the appropriate people, filled out and returned.

The original team that participated in the Vision Work Stage effort will most likely be expanded for the assessment activities. Here you will need to identify a large array of resources involved with managing and executing IT activities. These will become additional Extended and Advisor Stakeholders.

The overall assessment strategy and approach should also be reviewed. From this, you can determine if workshops will need to be scheduled along with key meetings with Senior Management personnel to review the results.

Once all of this is in place, it is recommended that an Assessment Stage Kickoff Session be held. This will serve to get everyone on the same page as to what is being done and the goals of the Assessment effort.

Assess Current State Processes Task

This task will involve participating in assessor workshops and interviews to determine the current state process maturity and gaps against the planned vision. The output from this task will be a report that lists the Process Assessment findings.

The approach for conducting this will primarily be driven by the outside parties you enlisted to conduct the assessment. During the workshops and interviews, the assessors will be comparing what they see and hear against ITSM best practices and the planned Vision.

Be aware that some staff may be nervous or suspicious about the assessment. The kickoff session can be one way to ally these fears. Also be aware of who participates together in workshop settings. Some staff may react defensive or not fully participate if they see their supervisors in the same room. For these situations it may help to hold separate workshops. I've always found it helpful, anyways, to find out individually if what management says is equal to what their staff says. I have seen many a situation where managers will proudly show me an IT policy that exists in some area and then find out that no one who reports to them is aware of it.

Assessment criteria for processes are still highly subjective when it comes to process maturity. The following high level guidelines may help you understand the general criteria that most assessors may look for when rating processes:

Example Process Maturity Scale

0	1	2	3	4	5
None	**Initial**	**Repeatable**	**Defined**	**Managed**	**Optimized**
Process not performed at all and no one is aware of what is needed	There is awareness, but processes are chaotic, ad-hoc and not documented	Basic processes are established and there is some level of adherence to them; some may be documented	Processes are well defined and documented; they are standardized and integrated with each other	Process quality is managed; data is collected on process quality and reported on a regular basis	Continuous improvement cycles occur with the process on a regular basis driven by business and technology changes

Assessors will typically have specific criteria at each individual process level to determine where your organization might fit into the above categories.

Assess Current State Technologies Task

In this task, the current state of support technologies is examined to determine how well existing technologies support IT Service Management best practices and the planned Vision. The focus of the technology assessment is on the general state of the employed tools and technologies and how well they will support the Vision. It is not a vendor tool comparison effort.

Typical findings in the technology assessment may be in areas such as:

- Missing key tools and technologies needed to support ITSM or the planned Vision

- Tools exist, but have not been implemented to their best advantage

- Utilization of many point tools that are not well integrated

- Tools that create more problems than they solve

- Use of redundant tools to do the same things (i.e. more than one Incident Management tool for example)

- Key tools and technologies that have no vendor support, custom built or out of warranty

A quick word here about custom built tools. This as a warning flag for most IT shops. Choosing to build and employ a custom tool should be the option of last resort. While many IT organizations do this to get special functionality or feel that they will bypass vendor costs, the long term costs and pain of this approach are seldom considered. By choosing this path, your staff is taking on the ongoing burden of keeping the solution current with technologies and supplying staff and training to take care of it over time. In many cases, IT organizations have put themselves in a tough position where a custom built tool is acting as a restriction to adapt to new business initiatives and plans.

The outcome for this assessment should yield generic recommendations for tool gaps that need to be overcome. It should not yield specific recommendations to purchase vendor tool products. It is also not within scope of this effort to do product comparisons.

A sample set of assessment criteria that outside assessors employ might look similar to the chart shown on the following pages:

IT Service Management Tool Assessment Criteria

Characteristic	0 None	1 Weak	2 Fair	3 Capable	4 Mature	5 World Class
Functionality	Missing	Provides little support for ITSM processes. Only works with some platforms.	Provides function that applies to some activities of ITSM processes. Only works with some platforms.	Provides functions that address most ITSM processes. Works with many needed platforms.	A large proportion of ITSM processes are addressed and all needed platforms are supported.	All ITSM process activities and platforms are fully supported.
Integration	None.	Integration with other ITSM tools only exists by providing flat files or other batch interfaces.	Integration among tools limited to common user interface and providing flat files or other batch interfaces.	Integration with some ITSM tools possible through included interfaces such as data sharing.	Effective and broad integration with other tools, either by documented or vendor-provided interfaces. Good cross-platform operations.	Shares data and cooperates with other tools to achieve management tasks transparently across different platforms.
Level of Automation	Manual activities only.	A significant proportion of ITSM processes receive no support from products.	A significant proportion of ITSM processes are user initiated, with little automated functionality provided.	Automation is reactive and works with operator direction.	Proactive functionality with considerable automation. Some exceptions are handled using policy definitions.	High degree of automation provided by the tool. Predictive and self-healing with full support for ITSM processes.

Characteristic	0 None	1 Weak	2 Fair	3 Capable	4 Mature	5 World Class
Usability	None.	Viewed as unusable or highly confusing. Probably due to implementation problems, process ambiguity, or environment incompatibility.	Usable only by the most highly skilled technical and senior staff.	Basic functions of the tool easy to use; advanced usage requires high skill level.	Most tool functions can be effectively used by those with low technical skill levels.	All tool functions can be used effectively by those with low technical skill levels.
Reporting	None.	Only standard reports.	Minor capability for user specified reports.	Tool provides standard reports. User can specify some customized reports. Some ad hoc and exception reporting capability for experienced users.	All reports can be customized. Customized reports can be added to standard report schedule. Ad hoc reporting for all users. Automated exception reporting.	Easy customization and scheduling of all reports. Ad hoc requests can be initiated automatically as well as manually. Automated exception reporting.
Data	None is produced.	Data is not kept after tool is used.	Data is written to a file. Data from multiple runs cannot be combined easily. No search capability.	Data added to cumulative database. Search on single keys possible.	Data kept in single database, frequently with data from other tools. Database may have multiple indices. Easy searching for specific data.	Data kept in a database with data from other management tools. Easy to find and combine data from various tools, time frames, etc.

| Characteristic | 0 | 1 | 2 | 3 | 4 | 5 |
	None	Weak	Fair	Capable	Mature	World Class
Communication	None.	Displays results on a console terminal only where results need to be manually viewed.	Interfaces to single variety of a few communication vehicles such as pagers, cellular phones, E-mail. Requires manual initiation.	Interfaces to multiple varieties of some communication vehicles. Some initiated automatically.	Interfaces to most communication vehicles. Automatic initiation.	Interfaces with standard communication vehicles automatically.
Openness	None.	Runs on a single platform only.	Runs in a single environment, has proprietary interface.	Adaptable to several environments or works with defined standard interfaces.	Adaptable to installed technology and desired standard interfaces.	Adaptable to a wide variety of technology.
Vendor Support	None.	Support provided only through specialized consultants not part of the vendor's organization. Vendor has declared the solution as non-strategic.	Support available on a limited basis. Vendor has warned that solution will not be supported in a future timeframe.	Solution is fully supported by the vendor but support is only provided during standard business hours.	Solution is fully provided by vendor at any time or day.	Solution is fully provided by vendor at any time or day. Vendor provides a variety of self-help, remote diagnostic facilities and training.

Assess Current State Organization Task

In this task, you will be analyzing the current state of your IT organization to assess how well it might support, or not support, a Service Management solution. Examples of the kinds of things to be sought after here include:

- Assessing what organizations would be impacted by the future state Service Management solution

- Assessing skill sets and skill capabilities to support the future state Service Management solution

- Assessing current roles and responsibilities to see what areas may be missing, overlapping or redundantly performed by several organizations

- Assessing the organization's readiness for change

- Assessing cultural barriers and gaps that will get in the way of your service vision

For identifying current roles and responsibilities, lay out each process and its key activities and then do an ARCI (Accountable, Responsible, Consults, and Informed) analysis against them. Look at the Work Breakdown structure presented earlier in this chapter for an example of this. Do not be surprised to find that some processes have no one responsible for them. You may also find that some processes have several organizations all claiming to hold accountability.

Another key task is the Stakeholder Analysis. Here you will be identifying key stakeholders who will be impacted by ITSM solutions by name. You will then be working to identify their wants and needs specific to each individual. This is a key step. It is here that you will find out who can support you, who might be neutral and who may be working against you. This is invaluable information to obtain as you start to build and flesh out your later implementation activities. See the chapter on Adapting to an ITSM Culture for more details on this.

I highly recommend using an outside vendor that provides Organizational Change assessments and services to assist you with much of the above. They typically come with a wealth of techniques and skills unique to helping organizations cope with behavioral and cultural changes. From experience, I have found these people invaluable in working with organizations in a structured manner to achieve this change. Remember that if the organization does not change, your solution will end up sitting on the shelf.

Assess Current State Governance Processes Task

In this task, you will be analyzing the current state of your service governance processes and establishing an ITSM service baseline. Typical areas to be looking for that relate to Governance include:

- Reports that describe the quality of IT processes

- Service reports

- IT statistics

- Service Improvement Programs

- Service Levels and/or Operational Targets

- IT Governance Processes and Controls

With this assessment task you are looking to see how mature the IT organization is in providing governance over the services it delivers. That is, how well IT measures, reports and takes action on the quality of the services it operates. An example of a maturity scale for Governance activities can be pictured as follows:

Example Governance Maturity Scale

0	1	2	3	4	5
None	**Initial**	**Repeatable**	**Defined**	**Managed**	**Optimized**
No real IT measurements are captured; IT does not report on service quality to customers; IT determines what, how and when services are delivered.	Some measurements and reporting exists, but these are used only by IT to check for problems; IT determines what, how and when services are delivered.	Measurements and reporting are done on a regular basis to check for service problems; little interaction with customers other than when issues occur.	Regular measurements and reports are produced that describe IT internal performance; some quality metrics exit; there is some customer interaction.	Regular measurements and reports are produced that describe service quality; there is some interaction with customers.	Regular measurements and reports are produced that describe service quality in customer terms; ongoing service improvement program is in place.

One of the outputs of the ITSM Vision Work Stage was the Program Success Metrics Table that describes which factors will determine the success of the overall ITSM effort. In that stage, we left the baseline values blank. An additional task is to now fill those in with current known values based on observations and information gathered in this work stage. In the event that a baseline metrics simply cannot be determined because no evidence exists, leave the factor marked as TBD. Do not be overly concerned with this. Improvement can be demonstrated by showing the ability later on to report on metrics that were not obtainable when you got started.

Analyze Assessment Findings Task

In this task, you will be assembling the findings of all the assessment activities. Key output includes:

- Maturity rating results

- Organization analyses for readiness, change and skill capabilities

- Inventory of findings

The Inventory of findings is built from analyzing results of assessments and interviews to determine gaps against the planned ITSM Service Vision. You may wish to present this inventory as shown in the example below:

Inventory of Assessment Findings Example

Process Findings		
Reference	Area	Finding Description
001	Problem Management	Comprehensive root cause analysis not performed.
002	Problem Management	No separation of incidents versus problems.
Technology Findings		
003	Integration	Most tools are standalone point solutions.
Organization Findings		
004	Capabilities	There are little or no ITSM skills among staff.
005	Cultural	Rewards and promotions are given to those who demonstrate the ability to pull things out of the fire when things go wrong.
Governance Findings		
006	Metrics	Metrics are only collected for IT analysis purposes and none exist that indicate quality of services being delivered to customers.

Expect that you will have many more findings than what is shown in the above example. The reference column is there because the findings will be cross referenced later with recommendations. The Area column helps sort out and group the findings in a logical manner. The Finding Description column describes the actual finding.

The above table also makes it much easier to justify your maturity scores. For example, if you rated Incident Management as a 1.0 (Initial), you can cite each finding that led to that decision by reference number. If there are a large number of findings, you may cite the top 3 or 4 as reasons for the rating.

Make sure you review the findings with a number of key stakeholders. It is important to make sure that there were no misunderstandings about what was taken from interviews and surveys. Update findings as necessary based on any feedback obtained.

Identify Recommendation Actions Task

For this task, you will be turning the findings developed earlier into action items. You will identify what things need to be done in order to close the gaps that were inventoried in your findings table. Action items are those steps that should be taken to move the IT organization towards the planned Service Vision.

At this stage, they are somewhat generic in nature. An example of bad action item at this stage would be:

Implement Vendor ABC's tool.

An example of a better action item would be:

Implement an Incident Management tool to automate tracking and reporting of incidents.

Similar to your findings, action items should also be inventoried. The table shown on the following page provides an example of how action items should be identified. It is recommended that a workshop be held with key stakeholders to do the scoring. When filling out the table, note that for some columns, the High Medium, Low designation is reversed.

If filled out and scored correctly, the higher the score number is, the higher the ranking is that the action item falls into a "quick win" or Initial Win category. This means that the action item represents an action that can be taken that can be done fairly quickly, at minimal cost and with few organizational barriers yet will yield a high amount of benefit.

If you are using an outside consultant to perform the assessments, you may want to make sure that their approach somewhat provides something similar. This approach ensures that there is a well audited back trail for the findings, conclusions and recommendations that you plan to proceed with.

Inventory of Action Items—Example

Ref = Action Item reference number.

Area = for process, list each ITSM process area. All others indicate Technology, Organization or Governance.

Action Item = describe the action item being recommended.

XRef = refer to the reference numbers for all findings being addressed by the action item.

Cost = indicate 1 (High), 2 (Medium) or 3 (Low)

Benefit = indicate 1 (Low), 2 (Medium) or 3 (High)

Effort = indicate 1 (High), 2 (Medium) or 3 (Low)

Barrier = indicate 1 (High), 2 (Medium) or 3 (Low)

Score = multiply each 1 in the row by 1, each 2 by 3 and each 3 by 9—then sum these across

Comments = indicate justification for action item and any additional comments as necessary

Ref	Area	Action Item	XRef	Cost	Benefit	Effort	Barrier	Score	Comments
001	Problem Management	Develop a root cause analysis workshop and conduct root cause analysis training and certification among IT staff.	001	3	2	3	3	87	Will contribute towards eliminating repeat incidents.
002	Technology	Implement an integrated suite of service support tools	002 003	1	3	1	2	35	Will provide efficient workflow for incidents and problems and reduce redundant labor tasks with these.
003	Organization	Implement employee bonus structure based on delivery of good service.	005	3	3	3	1	82	Needed to instill service mindset among IT staff and management.
004	Organization	Institute an ITSM skills training program.	004	2	3	3	3	87	Needed to instill service mindset among IT staff and management.

Identify Initial Win Efforts Task

For this task, you will be identifying the Initial Win efforts that you plan to proceed with. This should be a fairly straightforward task to complete if you filled out the Action Items Table described in the previous step. The approach to do this is as follows:

1. Pick a target value to identify which Action Items will be considered high priority (For example: all action items with score over 50 will be considered high priority)

2. Group like action items together

3. Form an Initial Win effort around each grouping

4. Add in any additional Action Items that fell into a less than High priority that would also be addressed by the Initial Win

5. Build a description for each Initial Win identified (see example on next page)

6. Estimate a timeline for each Initial Win (keep it simple—if the effort scored Low estimate 2-3 months, Medium 4-8 months, High 8-12 months)

7. Cross reference each Initial Win back to the Action Items it addresses

It is helpful to work through the above steps in a workshop setting. This increases the buy-in for the activities selected and provides stronger feedback for what Initial Wins will be selected, how they will be described and what the completion criteria is for them.

The following page shows an example of an Initial Win description. You will be developing a set of these for input to the presentations and materials that will be submitted to Senior Management personnel.

Initial Win Description Example

Initial Win	Build and publish a corporate wide Release policy.

Key Activities	Identify any exiting enterprise Release Policies
	Define and document a Release Policy
	Identify criteria for release categories, (i.e. Major, Minor, Emergency, etc.)
	Identify criteria for release types (i.e. Full, Delta, Package)
	Identify workflows for each release category and type defined
	Add criteria and workflows defined above to the Release Policy document
	Define a release numbering and naming convention standard
	Add numbering and naming convention standard to Release Policy document
	Identify Patch Management requirements
	Build Patch Management policy
	Incorporate Patch Management policy as part of Release Policy

Completion Criteria	This Initial Win will be considered complete when:
	A corporate policy has been approved by IT Senior Management
	A published policy exists on the Corporate Policy web site
	A training and certification program has been completed for IT staff

Estimating Factors	Duration: (Derive from the Action Items Inventory Table—i.e. 2-3 months)
	Cost: (Fill in from the Action Items Inventory Table)
	Benefit: (Fill in from the Action Items Inventory Table)
	Effort: (Fill in from the Action Items Inventory Table)
	Barriers: (Fill in from the Action Items Inventory Table)

Recommendations Addressed	(List here from your Inventory of Action Items Table)

Planned Benefits	(Use your Inventory of Action Items Table comments and Inventory of Findings to indicate planned benefits to be achieved by implementing this Action Item).

Justification	(Use your Inventory of Action Items Table and Inventory of Findings and results of the surveys done during the ITSM Vision Work Stage to indicate why this Action Item should be done now and what might happen if it were not done)

Obtain Go-Forward Agreement Task

This Work Task culminates in the Senior Management buy-in and go-ahead authorization. At this stage, the go-ahead authorization is to continue with the ITSM Planning Work Stage only. There will be additional meetings with Senior Management at the end of the ITSM Planning Work Stage before the final go-ahead is given to start.

The Work Steps within this Work Task will primarily focus on execution of the Vision Communications Strategy to shop the assessment findings and recommendations to key stakeholders leading up to the Senior Management meeting. This means executing those parts of the strategy deemed critical towards obtaining the buy-in to the next Work Stage. This will most likely involve presenting various versions of the findings to different stakeholders to make sure the appropriate decision makers and influencers are on board with the results.

You should also prepare a detailed project plan for the ITSM Planning effort (see the ITSM Planning Chapter for help). You may also need to prepare a Program Charter if your organization requires one.

Chapter
8

ITSM Planning Work Stage

*"There is a big difference between a million parts flying in
close formation at high speeds and an airplane"*
—A Wise Aviation Engineer

At this point in the transformation effort, the visioning and assessment efforts have been completed. You've established a future state service vision. You've assessed your current state and identified action items to get there. You've identified Initial Win projects and you're now ready to start implementing things in earnest. This chapter will get you going by helping you structure your plans and program operation to make sure your efforts succeed.

It is important to consider what needs to be in place in terms of Program operation before starting the transformation effort. The Program itself must be well organized. Certain project management basics need to be in place to make sure that a solid improvement foundation is underway and operating. Failure to do this may result in miscommunication among project teams, confusion over roles and responsibilities and lastly, a perception among teams and staff that the effort is not very well organized and doomed to failure.

The focus in this chapter is to arm you with information for structuring your program before implementation activities start. There will much in the way of Initial Win projects all running simultaneously. At the same time, strategic processes are being designed, policies are being built and more stakeholders are assisting in active roles to carry the vision. Jumping into all this without a solid program structure, working standards and detailed work plans will invite disaster.

At the end of this chapter, you should be armed with many of things you need to put together to ensure that your ITSM Transformation Program will operate with maximum efficiency and effectiveness. In short, you will have the roadmap needed to drive your transformation efforts forward.

ITSM Planning Work Stage Approach

If we look at the Planning effort over an approximate 4-5 week timeline, the above Work Tasks might be timed to look like the following:

A detailed Work Breakdown Structure is depicted on the following page:

ITSM Planning Stage—Work Breakdown Table

R=Responsible
A=Assists
C=Consults
I-Informed

Work Task	Work Step	Org	Proc	Tech	Gov	Mgt	Work Product
Manage Planning Work Stage	Manage Planning Work Stage activities	A	A	A	A	R	
	Report Planning status	A	A	A	A	R	Status Reports
	Coordinate and schedule Planning meetings	A	A	A	A	R	
	Integrate Planning activities across all work tracks	A	A	A	A	R	
Establish Program Office	Assemble Program Office staff	I	I	I	I	R	Assembled Program Office Staff
	Assemble Process Owners	C	C	C	C	R	Assembled Process Owners
	Assemble Program Teams	A	A	A	A	R	Assembled Program Teams
	Build Program Organization Structure	C	C	C	C	R	Program Organization Chart
	Assemble Program Baseline Metrics	A	A	A	R	A	Program Baseline Metrics
	Establish Work Products Acceptance and Review Procedure	C	C	C	C	R	Work Products Acceptance and Review Procedure
	Establish Email Distribution Lists	A	C	C	C	R	Email Distribution Lists

Work Task	Work Step	Org	Proc	Tech	Gov	Mgt	Work Product
Establish Program Working Standards	Establish Program Working Tool Standards	C	C	A	C	R	Program Working Tool Standards
	Establish Program Issues Log	C	R	C	C	A	Program Issues Log
	Establish Status Reporting Templates	C	R	C	C	I	Status Reporting Templates
	Establish Program Meeting Calendar	C	C	A	C	R	Program Meeting Calendar
	Establish Program Terminology Guide	C	A	C	R	C	Program Terminology Guide
	Establish Work Product Templates	A	A	A	R	A	Work Product Templates
	Establish Program Document Repository	C	C	A	R	C	Program Document Repository
	Establish Program Governance Process	C	C	C	R	C	Program Governance Process
	Establish Process Modeling and Documentation Standards	I	R	C	C	I	Process Modeling and Documentation Standards
	Develop and publish Program Working Standards Handbook	C	C	C	C	R	Program Working Standards Handbook
Prepare Program Work Plans	Develop detailed Initial Win Plans	A	R	A	A	I	
	Develop detailed Foundation Plans	A	R	A	A	C	
	Roll plans up into Program Master Plan	A	A	A	A	R	Program Master Plan
	Develop Work Product Delivery Map	A	A	A	A	R	Work Product Delivery Map
	Develop Program Risk Mitigation Strategy	A	A	A	A	R	Program Risk Mitigation Strategy
Prepare Project Teams	Develop Team Training Strategy	R	A	A	A	C	Team Training Strategy
	Obtain Team Security Clearances	I	I	C	I	R	Team Security Clearances
	Obtain Office Working Space	I	I	I	I	R	Office Work Space

Work Task	Work Step	Org	Proc	Tech	Gov	Mgt	Work Product
	Develop and Build Program Web Site	A	A	R	A	C	Program Web Site
	Develop Stakeholder Initiatives Inventory	R	C	C	C	C	Stakeholder Initiatives Inventory
Prepare Program Communications	Develop Stakeholder Map	R	A	A	A	A	Stakeholder Map
	Develop Resistance Plan	R	C	C	C	C	Resistance Plan
	Develop ITSM Overview Presentation	A	R	C	C	C	ITSM Overview Presentation
	Prepare Program Overview Presentation(s)	A	A	C	C	R	Program Overview Presentation(s)
	Prepare Kickoff Meeting Presentation	A	A	A	A	R	Kickoff Meeting Presentation
	Update cost estimates based on work plans	A	A	A	R	A	Updated Cost Estimates
	Update communication plans	R	C	C	C	C	Updated Communication Plans
	Draft management presentation(s)	R	A	A	A	A	Management Presentation(s)
Obtain Go-Ahead Decision	Schedule and conduct management presentation(s)	R	A	C	C	C	
	Update plans and assessments based on any management feedback	A	A	A	A	R	
	Obtain management approval to go forward with ITSM Foundation and Initial Wins efforts	A	C	C	C	R	Go-Ahead Decision

Planning Key Work Products

The table below presents each Work Product produced by the ITSM Planning Work Stage that is a milestone and provides a further description of it. Other Work Products are described further in the ITSM Planning Work Task Considerations section of this chapter.

ITSM Planning Key Work Products

Work Product	Description
Program Organization Chart	Chart of all participants to be involved in the implementation effort Participant contact information Includes all currently known Core, Extended and Advisor team personnel
Program Baseline Metrics	Collection of success factor metrics for the program as developed in the previous two Work Stages Presented as a Program dashboard report Metrics to be updated each month as the implementation proceeds
Work Product Acceptance and Review Procedure	Describes how Program Work Products will be reviewed, approved and signed off on Includes names of any key reviewers and approvers that will be involved with these tasks during the program effort
E-Mail Distribution Lists	Distribution lists for the Program along with procedures, roles and responsibilities for maintaining them
Program Working Tool Standards	Lists tools required create and maintain program work products (i.e. PowerPoint, Project, Excel, etc.) Identifies versions of tools to be used by the effort
Program Issues Log	Master log of program issues to be tracked and managed by the Program Office Includes inventories of the issues, who is involved with them and their current status
Status Reporting Templates	Identifies how implementation team members will report activities, issues and general status to the Program Office Identifies how team status reports will be rolled into an overall status report on the implementation effort Includes items such as activities performed last period, plans for following period, key issues and progress against the Program Master Plan
Program Meeting Calendar	Identifies all known Program meetings Allows team members to see who is meeting with who and for what purpose
Program Terminology Guide	Lists common Program terms and definitions. (i.e. what is meant by the word "process"). Should also include acronym definitions (i.e. CMDB = Configuration management Database).

Work Product	Description
Work Product Templates	Work Product standards for common Program look and feel Presentation templates with logos and formats Templates of key deliverables such as a Process Guide
Program Document Repository	Established repository to hold all Program working documents Authoritative source for all Program deliverables and documentation Includes procedures for accessing the repository as well as for updating it Separates documents that are work-in-progress from those posted for review or marked as final and approved Easy schema design to make it easy for Program team members to find things May include security controls over information that can be seen
Program Governance Process	Describes how Program scope will be controlled and managed Identifies how changes to the Program will be handled and escalated for decisions Identifies key responsibilities, roles and names of personnel who will be involved with Program Governance activities
Process Modeling and Documentation Standards	Describes format and tool standards Describes how processes are to be documented
Program Working Standards Handbook	Outlines general terms, conditions and responsibilities for how Program team members will work within the overall Program Published as a handbook Should be made available on the Program Web Site
Program Master Plan	Identifies key tasks, work products, milestones, timeframes and effort estimates for the implementation effort May include many sub-plans rolled into a master plan
Work Product Delivery Map	Lists every known Work Product to be produced by the Program Links Work Products to planned delivery dates and owners
Program Risk Mitigation Strategy	Lists all known Program Risks and associated mitigation actions to counter them
Team Training Strategy	Indicates actions to be taken to obtain appropriate level of ITSM skills for all team members Strategies may include everything from informal presentations on ITSM to formal classroom training and certification Indicates timeframes for when key training events will take place Identifies costs for training as necessary Identifies internal training materials to be produced if these will be used
Security Clearances	Security IDs, Passwords, procedures and clearances to get to items such as Program documents, Program repository, e-mail and calendars

Work Product	Description
Office Working Space	Working space for team employees such as cubicles, desks, PCs, chairs, phones, etc.
Program Web Site	Easily accessible web site that contains information about the program Contains announcements, accomplishments and key activities about the program
Stakeholder Initiatives Inventory	Lists outside initiatives to the Program that are planned or taking place that may be impacted, or have impact on, the overall Program (i.e. a Change Management process that is being developed by another organization) Identifies strategies for how to engage with each initiative
Stakeholder Map	Lists all Program Stakeholders along with their commitment levels, influence in the organization, and key concerns, wants and needs
Resistance Plan	Documents strategies for dealing with resistance to ITSM initiatives Inventories potential avenues for resistance behaviors
ITSM Overview Presentations	Used to inform stakeholders on ITSM basic concepts and understanding Not to exceed 1 hour in length May tailor different versions of these depending on audience (management, technical, etc.)
Program Overview Presentation(s)	Used to inform stakeholders on the implementation effort, general strategy and goals Not to exceed 1 hour in length May tailor different versions of these depending on audience (management, technical, etc.)
Kickoff Meeting Presentation	Program goals/approach/overview (steal from prior presentations) Participant Concerns and Issues Program Working Standards Planned Meetings and Next Steps

ITSM Planning Transition Team Organization

The recommended ITSM Planning Transition Team Organization is that pictured below:

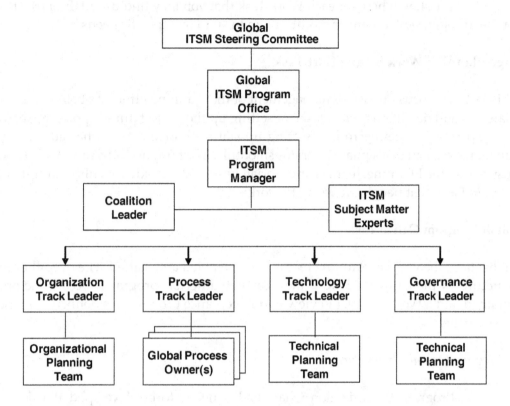

Similar to the prior Work Stages, the above recommended organization structure is a smaller subset of the large implementation organization structure (See ITSM Transition Organization chapter). This structure allows for the four Work Tracks: Organization, Process, Technology and Governance just as in the prior Work Stages. The teams that work within each track should consist of Core, Extended and Stakeholder types of individuals.

By organizing in this manner for planning activities, the Work Track Leaders greatly leverage the planning coordination that will occur. Without this, you will soon discover that many plans begin to overlap or have dependencies on one another. This leads to disorganized and frustrating planning sessions where large numbers of people are involved trying to sort through plans and dependencies that seem to go nowhere. It also avoids the situation where a single individual is given responsibility for plan coordination and usually ends up missing things or is perceived as a nuisance by other planners.

The Track Leaders will hold responsibility for building the plans that lie within their track. They will then coordinate with each other to manage the plan dependencies and roll plans up to a master plan under control of the Program Manager.

ITSM Planning Work Task Considerations

This section describes each of the ITSM Planning Work Tasks in a little more detail. Practical tips and ideas are shown here for each Work Task that you may find useful. Feel free to steal any of the ideas presented here or modify them to suit your specific needs.

Manage Planning Work Stage Effort Task

This task will focus on overall management of the Planning effort. As before, the Project Manager will do all the typical tasks of setting up the project, time reporting, obtaining work space and meeting rooms, setting up status reporting and other administrative functions to ensure this phase operates smoothly. Most importantly, this Work Task will ensure that the Planning Team is appropriately staffed, outside consulting resources are procured and that people are ready to work on it.

Establish Program Office Task

In this task, the various planning teams are assembled and staffed. The overall Program structure is then finalized. Key metrics that describe program success factors are obtained and a Work Product Review and Acceptance procedure is defined. To describe these further:

Program Organization Chart

A Program Organization Chart or list needs to be developed that identifies all known participants. This chart will be owned by the Program office. Key elements of the list include:

- Name
- Department
- Program Role (i.e. Core Team, Advisor Team, etc.)
- Location
- Contact information

Program Baseline Metrics

Program success metrics were initially developed during the ITSM Visioning Work Stage and updated in the ITSM Assessment Stage. In the ITSM Planning Stage, you will formalize these into a set of Program Metrics that can be monitored and reported on over time. These metrics will indicate whether the program is successful or not in achieving its objectives.

Reporting and tracking of metrics is a critical process within the transition program. When Senior Management starts asking tough questions like "How are we really doing?" or "What success have we started to see with all this effort we spending time on?" you will have a ready answer to show them.

Metrics will first be base lined and then updated on a periodic basis. Think of this document as a dashboard for the Steering Committee to indicate overall program progress towards its objectives. For the ITSM Planning Stage, the document should be developed and populated with the baseline values. Subsequent versions should show how those values have changed as the program progresses.

Work Product Acceptance and Review Procedure

A procedure should be documented that describes how Program Work Products will be reviewed, approved and signed off on. This should also include the names of any key reviewers and approvers that will be involved with these tasks during the program effort.

The procedure should be included in the Program Working Standards Handbook.

E-Mail Distribution Lists

Sets of Email distribution lists should be established for the Program along with procedures, roles and responsibilities for maintaining them. Examples of the kinds of lists to develop might include:

- Everyone
- Core Team Participants
- Steering Group
- Extended/Advisor Team Participants
- Process Owners
- Process Owners and Stakeholders
- Any other groupings as necessary

Having these lists in advance greatly enhances overall project communications. It also ensures that participants do not get inadvertently left out for key communications.

Establish Program Working Standards Task

In this task, all the working standards to be used by the implementation team are defined. These set up the guidelines for how the team will operate, communicate and provide status throughout the effort. To describe these further:

Program Working Tool Standards

A working standard needs to be identified for all the Work Products that will be produced in the transformation program. This standard should identify the tools and templates to create and maintain them. Examples of these might include which word processing tool should be used for reports or which presentation tool for creating presentations.

You should be fairly specific in identifying products and version levels to avoid the situation where some documents produced by some team members cannot be read by others. One way for documenting this is to create a list of the Work Products to be produced in a table and have the appropriate tools and versions listed alongside them for easy reference.

Other working standards might include names of libraries where Work Products will be stored. It might also include procedures for how Work Products will be considered complete and final and migrated to these locations. Templates may also be created for standard items such as Status Reports or backgrounds and logos for presentations.

Program Issues Log

To ensure that nothing falls through the cracks as the program progresses, a master log is kept by the Program Office that documents program issues when they arise and their current status. Elements of this log should include:

- Program Issue Descriptive Title
- Issue Status Category (Open, Pending, Closed, etc.)
- Issue Owner Name
- Description
- Priority (High, Medium, Low)
- Issue Open Date
- Current Issue Status Description

Responsibility for the Issues Log will also lie with the Program Office. This log should also become part of any periodic status reporting that takes place within the program. The Program Office needs to maintain and monitor this log to ensure that program issues are addressed in a timely manner to avoid unplanned program delays.

Status Reporting Template

Templates should be developed for reporting progress as the program executes. There will be two kinds of templates. One template will be used to present status from each Work Track to the Program Office. The other template will be used to present status from the Program Office to the Steering Committee. This latter template will cull information from each Work Track to present overall program status.

It is recommended that Program status be reported on a weekly basis. Key elements of the templates should include:

- Status working period (from date, to date)
- Progress against the Program Master Plan
- Planned Progress for the following period
- Key issues and concerns
- Copy of Program Issues Log with action updates

It will be the responsibility of each Work Track to document and present status to the Program Office. It will then be the responsibility of the Program Office to produce the final Status Report that will be presented to the Steering Committee. The Program Office will also store and maintain each status report.

Program Meeting Calendar

A calendar should be developed that identifies all known Program meetings. This should be kept up to date at all times. The calendar should be accessible by all team members at any time to check on meetings that have been scheduled. It can also be used as a reference to check on available times that might be open for scheduling meetings as the need occurs.

The calendar will be a critical communication item for Program teams. It is important that each Program team member not only view the items that concern them individually, but that they also see which participants are involved in meetings and for what purpose.

Types of calendar entries might include:

- Status Meetings
- Working Sessions
- Training Sessions
- Road Show Presentations
- Any meeting that a team member wishes to hold with someone

Program Terminology Guide

As the program proceeds, you will find that terms and acronyms will arise in discussions among team members. These may not be readily understood by team peers or newer members joining the program. There may also be instances where certain terms mean different things to different people. Examples of this may be terms like "process" or "infrastructure".

Top counter potential confusion around this, it is helpful to build a Terminology Guide that lists common Program terms and definitions. The guide itself should be structured like a dictionary. Terms should be placed in alphabetical order with their definitions.

A separate section should specify acronym definitions such as CMDB (Configuration Management Database) or ITSM (IT Service Management). The acronym section just needs to explain what the term stands for, not document what it is. For example, this section might state CMDB = Configuration management Database, but not get into describing what a CMDB is or how it is used.

The Program Terminology Guide should be owned and maintained by the Organization Track. A procedure should be put into place for adding, modifying or deleting items in this guide.

Work Product Templates

To create a common look and feel for all Program deliverables and presentations, templates may be devised. Examples of these might include:

- Presentation templates with logos and formats
- Templates of key deliverables such as a Process Guide

Responsibility for developing, storing and maintaining templates will lie with the Program Office.

Program Document Repository

A repository will need to be established to hold all Program working documents. This ideally should be centralized. It should be recognized as the authoritative source for all Program deliverables and documentation. Procedures will need to be developed for accessing the repository as well as for updating it. These will become part of the Program Working Standards.

The repository itself should have separation between documents that are work-in-progress, posted for review and marked as final and approved. It should be designed with a schema that makes it easy for Program team members to find things.

Security aspects of the repository should also be considered. It may be desired that not all items should be seen by all (such as the Stakeholder Map). There may be special needs to allow non-local team members to access the repository.

Program Governance Process

Program Governance processes will be developed, documented and maintained by the Governance Track. This needs to be done ahead of time to ensure that project controls over scope and development are in place from the start. The Governance processes will also identify how changes to the Implementation Program will be handled and escalated for decisions. It also identifies key responsibilities, roles and names of personnel who will be involved with Program Governance activities.

The Governance Track will pull this document together and submit a copy to the Program Office. Key processes and procedures around Governance will be added to the Program Working Standards Handbook.

Process Modeling and Documentation Standards

Documentation formats and tool standards needs to be developed for how processes and working Procedures are to be modeled and documented. Examples of these might include a choice of:

- IDEF0
- LoveM Diagrams
- Flowcharts
- Swim Lanes
- Other standards that your organization may wish to use

Development of these standards will be the responsibility of the Process Track with some assistance from the Technical Track if modeling tools will be used.

Program Working Standards Handbook

This handbook outlines general terms, conditions and responsibilities for how team members will work within the overall Implementation Program. Items for this might include:

- When status reports are due

- How to report time/labor

- Account codes to use for time reporting or expenses

- Who produces status reports/information

- Use of the Program Calendar

- Who/where the definitive source for Program messages is/comes from

- How Key Program messages and work products will be communicated

- How teams will meet/communicate given geographic differences

- Meeting and meeting room scheduling procedure

- Who is responsible for key pieces of the overall Program

- Where to go for Program questions, issues and concerns

- Program Governance roles and responsibilities

- Use of Program document repository

- Program security procedures to access Program documentation

- Program distribution lists and use of E-Mail

- Program deliverables review procedure

- Key tools and templates needed by Program participants

- Other items deemed necessary for program operation

Having these items documented in a single handbook makes it much easier to orient new participants as they join the program. It also ensures that program operation occurs smoothly with a minimum of confusion among participants.

Prepare Program Work Plans Task

This task covers the transition work plan development effort. This should start with each Work Track developing its own implementation plans. These are then rolled up to a Master Plan owned by the Program Office. To describe these further:

Program Master Plan

When developed, a working version of the Program Work Plan should be finalized as the Master Plan. The Program Master Plan is basically a work plan document that includes tasks, roles, effort estimates, milestones and key timeframes. It can be built starting with the Transition Program Work Breakdown Structure described in the ITSM Design and Initial Wins chapters. For larger efforts, it is not unusual to have many sub plans that will roll up into a single Master Plan.

Responsibility for the Master Plan will lie with the Program Office. This includes maintenance tasks to keep the plan up to date when changes are introduced. The Master Plan will be the keystone by which the Program Office will manage and oversee the entire transition effort.

Production of the Master plan will start with each Work Track. The Work Track Leader will have responsibility to coordinate activities and plans with the other

Work Track Leaders. The Program Office will have final oversight to make sure the plans are complete and ready to go.

Work Product Delivery Map

This document should list every known Work Product to be produced by the Program and link them to planned delivery dates and owning Work Tracks. Having an inventory of all the Program Work Products makes it much easier to track overall program progress as the program executes. It also serves a reference to each work track top ensure Work Products are being delivered on a timely basis.

Key elements of this might include:

- Work Product Name
- Owning Work Track
- Owning Contact Name
- Final Approver Contact Name
- Status (Not Started, In Progress, Completed, Late, etc.)
- Work Product Description
- Draft and Final Due Dates

Program Risk Mitigation Strategy

Before transition activities start, it is important to make sure everyone is aware of the potential risks that exist that threaten the success of the transition effort. This document should list all known Program Risks and then describe associated mitigation actions to counter them.

An example of a risk might be something such as "Poor process architecture skills exist within the team". An associated mitigation action might be "Hire process architecture consultant" or "Train personnel in process architecture concepts".

Each risk should be listed along with its mitigation action to counter it. Key elements of each item in the list should include:

- Risk Title
- Risk Severity (High, Medium, Low)
- Risk Owner
- Risk Description
- Risk Impact if not addressed
- Risk Mitigation Action

It is the responsibility of the Program Office to pull this strategy document together using expertise from each of the Work Tracks. This document should also be updated over time by the Program Office as new risks are discovered or mitigation actions change.

Prepare Project Teams Task

This task covers several preparation tasks to get project teams ready to start transition efforts. Key items include:

Team Training Strategy

Each participant in the transition program should be reviewed as to their current level and understanding of ITSM concepts within the role they will play. For example:

• Process Owners will need to fully understand the best practices for the processes they represent

• Stakeholders may need only a basic level of knowledge of ITSM concepts.

A strategy should be developed for each participant to address how the appropriate skills will put into place. This might mean identifying ITSM training courses for example. Other examples might include developing overview presentations, certifying some participants and developing reading lists of materials.

The Organization Track will hold the responsibility for identifying the skill gaps in the team, putting a strategy into place to address them and then ensuring that the strategy is executed on. This will also involve training administration on the part of the Organization Team to monitor who has taken which courses and ensure that skills are being developed as planned.

Security Clearances

Security procedures to get to Program documents, repository, e-mail and calendars should be established and documented. These should be included with the Program Working Standards. Security clearances, badges, user IDs and access codes should also be obtained for any consultants or outside personnel working with teams.

Responsibility for obtaining Security Clearances will lie with the Program Office with technical assistance from the Technology Track.

Office Working Space

Although it may seem obvious, office working space and locations need to be obtained and confirmed for implementation team members. This may involve relocating staff to a common work location or establishing a virtual working environment by way of the company intranet or external internet. Lack of attention to this could result in poor team morale and introduce delays into the implementation schedule.

Key considerations should include:

- Working space such as cubicles and offices
- Desks, chairs and file cabinets for implementation staff
- Meeting room locations
- PCs, phones and printers for staff
- Bridge or conference lines for phone meetings
- Web conferencing facilities if these will be used

Responsibility for establishing Office Work Space facilities will lie with the Program Office.

Prepare Program Communications Task

In this tasks, parts of the communication strategy developed in the prior Work Stages is executed to ramp up project teams and prepare for implementation activities. Much emphasis is spent on reviewing and updating the stakeholder lists developed in the prior Work Stages. These are refined to build a stakeholder map. Additional stakeholders may be added in Extended and Advisor team roles.

Key items produced by this task area:

Program Web Site

A key communication activity is to build and manage the Program web site. This is a great way to communicate Program activities and accomplishments as the transition effort proceeds. An example is shown on the next page:

Example of an ITSM Program Web Site

ITIL Support

Sub-Processes

In Flight
  Change Management
  Configuration Management

Ready for Kick off
  Service-Level Management

Awaiting Kick off
  IT Continuity
  Capacity Management
  Problem Management
  Security Management
  Financial Management

A Message From Your Service Manager

The work that we've done this year has been tremendous and, it's a testament to the dedication of the team, the commitment of the team, the talent and skills that they bring, and just the pride that you have in what you do every day, it comes through. Great Job!!!

Look At Our Accomplishments!!!

•Reduced outages by 20%
•Established liaisons with our units
•Implemented the new Service Desk
•Increased Customer Satisfaction Rating by 30%
•87 Employees ITIL Certified!
•Many others – check here!

Employee IT Service Award

ITIL Information

Activities Spotlight

Types of items to put on this web site might include item such as:

- Program Announcements
- Program Staff Profiles
- Program Accomplishments
- Program Progress Information
- ITSM Overview
- Information Specific to Key Program Teams
- Awards
- Certification Information
- Service Management Training Information
- Upcoming Service Management Events
- Other Items Of Interest

Stakeholder Initiatives Inventory

This document lists outside initiatives to the Program that are planned or taking place that may be impacted, or have impact on, the overall Program. An example of this might be a Change Management process that is being developed by another organization.

It is important to make sure that all of these initiatives are identified along with a strategy for how to engage with them. The engagement approach might include options such as actively working with the outside initiative, making a contact part of the Stakeholder Team or simply working only on a need-to-know basis.

Key elements of the inventory should include:

- Initiative Name
- Initiative Sponsoring Business Unit
- Initiative Key Contact
- Initiative General Description (Key objectives, scope, target dates, etc.)
- Engagement Approach (Ignore, Make Part of Team, Put on Advisor Team, Attend some/all meetings, etc.)

The Stakeholder Initiatives Inventory should be developed and maintained by the Organization Track. A copy of this should be stored with the Program Office.

Stakeholder Map

In prior Work Stages, key stakeholders were sought out and identified. One of the key tasks to be performed by the Organization Track is to formalize the stakeholder information collected earlier and document it. This document will list all Program Stakeholders along with their commitment levels, influence in the organization, and key concerns, wants and needs.

Key elements your Stakeholder Map should include:

- Stakeholder Name
- Organization Title
- Department/Location
- Level of Influence (Major, Minor, etc.)
- Needs (What's in it for them?)
- Concerns/Issues
- Program Opinion (Champion, Follower, Resister, Neutral, etc.)
- Level of Commitment (i.e. 2 hours per week, full time, etc.)

Resistance Plan

Having done the stakeholder mapping effort, you may discover that some staff may be actually blocking or resisting the ITSM effort. If this rears its head, you may wish to develop some strategies in advance to cope with these stakeholders as the effort gets underway. Some examples of strategies to use might include:

- Establish a series of meetings to resolve conflicts
- Negotiate program benefits or scope that might get a needed stakeholder on board
- Explore further communication options

- Find ways to neutralize the resistor so that he/she does not hinder program progress
- Ensure that resistance is not due to miscommunication

Having a resistance plan in advance will allow you to quickly cope with those that might sabotage your efforts. It allows you to raise awareness when this happens and not get caught off guard in a manner that might hinder program progress.

ITSM Overview Presentation

You will soon discover that this is one of several presentations that you literally need to keep in a Program Office handbag because it is given many times to different audiences. It is needed to explain basic IT Service Management concepts and why they are being used with the Program.

You may need to develop several versions of these depending on the audience you address. For example, it may go into more depth for IT personnel, less depth for a manager or business unit person. It should be no more than 1 hour in length. It should get right to the main concepts and key points.

We are not trying to make ITSM experts with this presentation. The main goal is to introduce stakeholders and others joining the program about the basic concepts so that there is a minimum understanding as to why things are being done the way they are. It is assumed that audiences receiving this presentation have little or no knowledge about ITSM.

The responsibility for producing this presentation should be owned by the Process Track. The Organization Track may assist in terms of making sure concepts are clearly stated and that the level of presentation is appropriate for the audience it will be delivered to.

Program Overview Presentation(s)

This is another kind of presentation that you will find needs to be given over and over incessantly throughout the transition program. This presentation will be used to explain the goals of the transition program, its objectives and overall approach. This should be at high levels and not go into great detail.

This presentation will be used for introducing new stakeholders and other participants to the transition program. It should let them know what the transition program is all about as well as how it is structured and operated. The presentation itself should not be more than 1 hour in length.

The responsibility for producing this presentation should be owned by the Program Office. The Organization Track may assist in terms of making sure concepts are clearly stated and that the level of presentation is appropriate for the audience it will be delivered to.

Kickoff Meeting Presentation

A presentation should be developed to cover the Program Kickoff Meeting Work Task that will occur at the beginning of the Design Work Stage. The Kickoff Meeting essentially fires the starting gun on implementation activities. The very first meeting should be done with at least all the Process Owners.

Much of the content for this presentation will be culled from other presentations described earlier. Key elements that will be included are:

- Program goals/approach/overview (steal from prior presentations)
- Participants Concerns and Issues
- Program Working Standards
- Planned Meetings and Next Steps

Responsibility for creating this presentation will lie with the Program Office with assistance and input from all the other work tracks.

Obtain Go-Ahead Decision Task

Here is the key task for getting that final decision and approval to move forward. The Work Steps within this Work Task will primarily focus on execution of the Vision Communications Strategy to communicate the final costs, approach, risks and timeframes for the effort. As before, you may end up doing this through a series of stakeholders leading up to the Senior Management meeting.

Chapter
9

ITSM Design Work Stage

> "Great companies cannot be built on processes alone. But believe me, if your
> company has antiquated, disconnected, slow-moving processes—particularly
> those that drive success in your industry—you will end up a loser"
> —Louis V. Gerstner, Jr. CEO, IBM

The ITSM Design Work Stage is where design efforts will take place to build the strategic future state ITSM solutions. It is in this stage where you will be building your future state process elements such your process mission, key activities, policies, procedures, tooling and organization strategies.

Note that this Work Stage is focused on building the strategic ITSM solution. This differs from the Initial Win Work Stages which are dedicated to building and implementing tactical solutions. If we take one of the processes, such as Service Level management for example, the Design Work Stage would focus on identifying the future state Service Level Management process, tooling strategy and organization. The Initial Win Work Stages would focus on specific solutions such as implementing a Service Catalog, Operational Level Agreements and Service Level Agreements in a manner that allows these to be in place within several months.

The ITSM Design and Initial Wins Work Stages really work hand-in-hand together for building solutions. The Initial Wins Work Stage will be used to implement key aspects of your ITSM solution that have been identified as having the most benefit and shortest implementation effort. The Design Work Stage will serve to guide and steer how those win solutions are put into place to ensure that all solutions being implemented are in accordance with strategic plans and direction.

This chapter presents a very prescriptive approach for designing and building process solutions step by step. At the end of this chapter you should have a pretty solid idea of how to go about building and integrating all the elements that go into your future state ITSM solution.

ITSM Design Work Stage Approach

If we look at the Design effort over an approximate 12 week timeline, the key Work Tasks might be timed to look like the following:

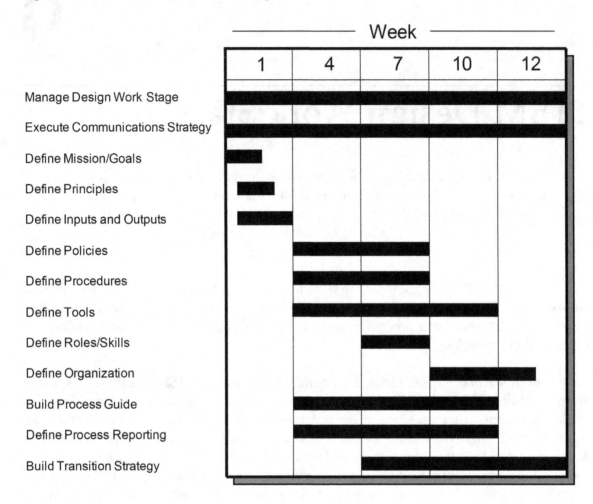

A detailed Work Breakdown Structure is shown as follows:

ITSM Design Work Stage—Work Breakdown Table

R=Responsible
A=Assists
C=Consults
I-Informed

Work Task	Work Step	Org	Proc	Tech	Gov	Mgt	Work Product
	Manage Foundation Work Stage activities	A	A	A	A	R	
	Report Foundation status	A	A	A	A	R	Status Reports
Manage Foundation Work Stage	Coordinate and schedule Foundation meetings	A	A	A	A	R	
	Integrate Foundation activities with Initial Wins as needed	A	A	A	A	R	
	Integrate Process and Transition plans across all processes	A	A	A	A	R	
	Review communications plan	R	A	A	A	A	
	Execute communications activities	R	A	A	A	A	
	Identify additional stakeholders	R	A	A	A	A	Updated stakeholder lists
	Perform stakeholder management	R	A	A	A	A	
Execute Communications Strategy	Obtain consensus on process decisions	R	A	A	A	A	Agreed Process Solutions
	Review communications plan	R	A	A	A	A	
	Execute communications activities	R	A	A	A	A	
	Identify additional process stakeholders and initiatives	R	A	A	A	A	Updated stakeholder lists
	Schedule and conduct Vision Stage Kickoff Session	R	A	A	A	I	

Work Task	Work Step	Org	Proc	Tech	Gov	Mgt	Work Product
Define Mission/Goals	Draft Process Mission Statement	A	R	C	C	I	Process Mission Descriptions
	Draft Process Goals	A	R	C	C	I	Process Goals
	Draft Critical Success Factors (CSFs)	A	R	C	C	I	Process CSFs
	Draft Key Performance Indicators (KPIs)	A	R	C	C	I	Process KPIs
	Define Process Scope	A	R	A	A	C	Process Statement Of Scope
Define Principles	Draft Process Guiding Principles	A	R	A	A	I	Process Guiding Principles
	Draft Principle Rationales	A	R	A	A	I	
	Draft Principle Implications	A	R	A	A	I	
	Identify additional Overriding Principles if needed	A	R	A	A	I	Updated Overriding Principles
	Integrate principles with other processes as needed	C	C	C	R	I	Updated process principles
Define Inputs and Outputs	Review High Level Process Activities	C	R	C	C	I	
	Review High Level Process Inputs	C	R	C	C	I	
	Review High Level Process Outputs	C	R	C	C	I	
	Draft Process SIPOC	C	R	C	C	I	Process Inputs and Outputs
Define Policies	Identify process policies	A	R	A	A	I	Policy Descriptions
	Draft individual process policies	A	A	A	R	I	Process Policies
	Draft ITSM Policy Guide	A	A	A	R	I	ITSM Policy Guide
	Identify Process Workflows	C	R	C	C	I	Process Workflows
	Draft Procedures	C	R	C	C	I	Procedures
Define Procedures	Identify frequency for each procedure	C	R	C	C	C	Procedure Frequencies
	Draft Work Instructions where needed	A	R	A	A	I	Work Instructions

Work Task	Work Step	Org	Proc	Tech	Gov	Mgt	Work Product
Define Tools	Review High Level Architecture	C	A	R	C	I	
	Draft Tool Requirements	C	A	R	C	I	Tool Requirements
	Review Existing Tools Inventory	C	A	R	C	I	
	Map Existing Tools Against Requirements	C	A	R	C	I	
	Identify Tooling Gaps	C	A	R	C	I	Tooling Gaps
	Define Tooling Direction	C	A	R	C	I	Process Support Architecture
	Integrate tooling strategy with overall architecture	I	C	R	I	I	Updated Tooling Architecture
Define Roles/Skills	Review High Level Roles	R	A	A	A	I	Process Role Descriptions
	Review Skill Requirements	R	A	A	A	I	Skills List
	Define Role Activities	R	A	A	A	I	Role Activities
	Define skills for each role	R	A	A	A	I	Skill/Role Maps
Define Organization	Map into existing organization	R	C	C	C	C	
	Identify skill/resource gaps	R	C	C	C	C	Sill/Resource Gaps
	Define ITSM organization	R	C	C	C	C	Future State ITSM Organization
	Define transition path to organization	R	C	C	C	C	Organization Transition Strategy
	Document Service Organization Strategy	R	C	C	C	C	Service Organization Strategy
	Identify training strategy	R	C	C	C	C	Training Strategy
Build Process Guide	Define Process Guide Template	C	R	C	C	C	Process Guide Template
	Draft Process Guide	C	R	C	C	I	Process Guide
	Obtain guide feedback	C	R	C	C	I	Updated Process Guide

Work Task	Work Step	Org	Proc	Tech	Gov	Mgt	Work Product
	Identify process reporting metrics	C	A	A	R	I	Inventory of Reporting Metrics
	Identify calculations and assumptions	C	A	A	R	I	Metric Calculations and Assumptions
Define Process Reporting	Identify data sources and repositories	C	A	A	R	I	Metric Data Sources
	Define reporting templates	C	R	C	C	I	Process Reporting Templates
	Identify tool support strategy	C	A	A	R	I	Metric Tool Strategy
	Identify reporting procedures	C	A	A	R	I	Reporting Procedures
	Implement baseline for process	C	A	A	R	I	Metrics Baseline
	Integrate reporting procedures with other process reporting	C	C	C	R	I	Integrated Reporting Procedures
	Identify early launch strategy	A	R	A	A	I	Early Launch Strategy
	Identify gaps between process and current state	A	R	A	A	I	Process Gaps
Build Transition Strategy	Draft process transition plan	A	R	A	A	C	Process Transition Plan
	Integrate process transition plan with ITSM plans	A	A	A	A	R	Integrated ITSM Transition Plan
	Integrate Foundation Work Products With Initial Wins	A	A	A	R	A	

Design Key Work Products

The table below presents each Work Product produced by the ITSM Design Work Stage that is a milestone and provides a further description of it. Other Work Products are described further in the ITSM Design Work Task Considerations section of this chapter.

ITSM Design Key Work Products

Work Product	Description
Process Mission Descriptions	Statement of mission of each ITSM Process area Includes inventory of goals for each ITSM Process area
Critical Success Factors (CSFs)	Inventory of performance requirements of a process that indicate it is operating successfully from a customer perspective Must be measurable with KPIs (Key Performance Indicators) Separate inventories exist for each of the ITSM process areas
Key Performance Indicators (KPIs)	Inventory of measurement metrics that indicate one or more Critical Success Factors are being met Established metrics should: Be measurable Establish a target and acceptance range Align to one or more Critical Success Factors
Process Statement Of Scope	Indicates the overall scope and impact covered by the Process Areas to consider might include: Geographies affected Organizational Units impacted Customer groups impacted Lines of Business affected
Process Guiding Principles	Statements of direction about scope, behavior and management philosophies that will influence how ITSM solutions will be designed Some Principles will be process specific, others may be applicable across all processes Key elements include: Principle Statement Example of how the Principle is used Rationale for why the Principle is in place Implications for working with the Principle
ITSM Policy Guide	Describes specific policies for each ITSM Process area at the procedural level Any other related policies A separate guide should be developed for each ITSM Process area and rolled up into an overall ITSM Policy Guide
Process Workflows	Describe Process key activities, inputs and outputs May include workflow diagrams for each ITSM Process area plus an integrated diagram covering all process areas

Work Product	Description
Process Mission Descriptions	Statement of mission of each ITSM Process area Includes inventory of goals for each ITSM Process area
Critical Success Factors (CSFs)	Inventory of performance requirements of a process that indicate it is operating successfully from a customer perspective Must be measurable with KPIs (Key Performance Indicators) Separate inventories exist for each of the ITSM process areas
Process Role Descriptions	Inventory of each role needed in the execution of the ITSM Processes Separate inventories should be developed for each ITSM Process area Includes a definition and description of each role Should list typical activities to be executed by each role
Skill/Role Map	Matrix of ITSM Process Roles and their level of responsibility Maps roles against organizational entities May describe this with RACI indicators: Responsible Assists Responsible Informed about activities only
Service Organization Strategy	Describes an organizational strategy to be implemented for executing the ITSM Processes Indicates a mapping of Process Roles to organizational job structure Describes an organizational implementation and transition strategy to organize around ITSM delivery of IT Services
Training Strategy	Identifies skills gaps that exist in current organization Identifies strategy to address skills gaps Lists planned training approach and delivery Indicates roles and responsibilities for training Indicates how confirmation of knowledge transfer will take place Includes cost estimates for training activities
Process Guides	Separate guide for each ITSM process area that includes all of the above work products A separate guide should be developed for each ITSM Process area
Process Reporting Templates	Templates and mockups of service reports for each ITSM Process area Some reports may be specific for each ITSM process area, others may span across several or all of the process areas
Integrated Reporting Procedures	Documents procedures for reporting on process quality Indicates how metrics will be extracted, calculated and reported Lists roles and responsibilities for reporting Indicates tools to be used for reporting
Early Launch Strategy	Documented strategy for how ITSM services will be tested with live business organizations prior to rollout Identifies target organizations to receive early launch services Identifies key timeframes and milestones for early launch activities Identifies roles and responsibilities for early launch activities

Work Product	Description
Process Mission Descriptions	Statement of mission of each ITSM Process area Includes inventory of goals for each ITSM Process area
Critical Success Factors (CSFs)	Inventory of performance requirements of a process that indicate it is operating successfully from a customer perspective Must be measurable with KPIs (Key Performance Indicators) Separate inventories exist for each of the ITSM process areas
Integrated ITSM Transition Plan	Documented strategy for how ITSM activities will be rolled out to business organizations Identifies key timeframes and milestones for rollout of all ITSM services Identifies roles and responsibilities for rollout activities

ITSM Design Implementation Team Organization

The recommended ITSM Design Implementation Team Organization is that pictured below:

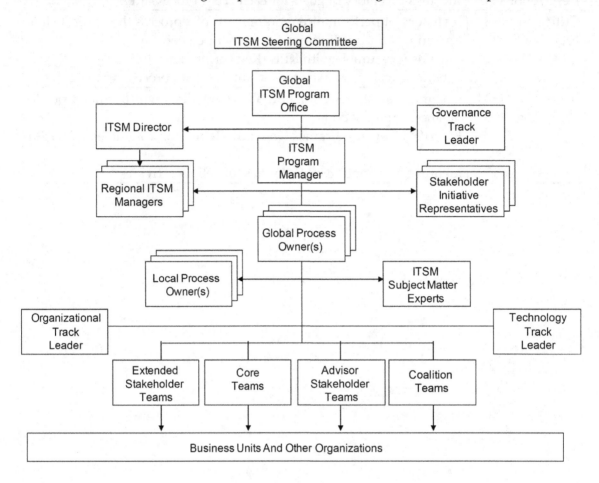

The above Program Implementation Organization structure is discussed in length in the Transition Organization Chapter earlier in this book.

ITSM Design Work Task Considerations

This section describes each of the ITSM Design Work Tasks in a little more detail. Practical tips and ideas are shown here for each Work Task that you may find useful. Feel free to steal any of the ideas presented here or modify them to suit your specific needs.

Manage Design Work Stage Task

This task will focus on overall management of the design effort. As before, the Project Manager will do all the typical tasks of setting up the project, time reporting, obtaining work space and meeting rooms, setting up status reporting and other administrative functions to ensure this phase operates smoothly. Most importantly, this Work Task will ensure that the design team is appropriately staffed, outside consulting resources are procured and that people are ready to work on it.

Execute Communications Strategy Task

In this task, the Communications Strategy developed in the prior Work Stages is now administered and carried out by the Organization Team. Depending on the tasks and schedules developed in that strategy, the Organization Team will perform all the needed functions to obtain buy-in for ITSM solutions and communicate ITSM activities across the organization and its stakeholders. The specific tasks to be performed will be driven from the Communications Strategy itself.

More detail around building and managing the Communications Strategy can be found in the Adapting to an ITSM Culture Chapter in this book.

Define Mission/Goals Task

The first step in defining and formalizing the ITSM processes is to identify a primary mission, goals, objectives and scope for each one. It is recommended that this be done by the Process Owner, related process Core Team and selected Extended Team stakeholders.

A Workshop should be held to identify these items. Approaches that can be used to define them include:

Mission Statement and Goals

ITSMLib provides examples of these for each ITSM process within the example process guides in the library. The approach for brainstorming the Mission Statement can be similar to that used for the Vision Statement in the ITSM Visioning Work Stage.

Critical Success Factors (CSFs) and Key Performance Indicators (KPIs)

Critical Success Factors identify what things must be done to ensure the process is operating successfully. Key Performance Indicators represent specific metrics that will tie into one or more Critical Success Factors to see if it is being met. ITSMLib provides many examples of metrics (KPIs and CSFs) that can be used.

Process Statement of Scope

This just represents a brief description that indicates who or what the process will be applied to within the business organization. This can be areas such as:

- Geographies
- IT Groups or Departments
- Lines of Business
- Groups and Users of Specified Technologies
- Other Organizational Units in the company
- Certain Customers or Customer Groups

Some examples might be as follows:

- The scope of this process will include all customers and lines of business served by our North American Data Centers.

- The scope of this process will cover all IT services provided to our Commercial Lending business unit.

- The scope of this process will cover all IT services delivered via our mid-range platforms and technologies.

Define Principles Task

This task covers the development of Guiding Principles unique to each ITSM process area. In the Visioning Work Stage, Overriding Principles were drafted. These covered Guiding Principles that are common across all ITSM processes. In this task, the effort is similar, but you will be focusing more on those specific and unique to each process itself.

Process Specific Guiding Principles

The outcome of this effort will yield process specific Guiding Principles. These are unique to each process. Examples of these can be found throughout ITSMLib.

Updates to Overriding Guiding Principles

This represents changes and additions to the Overriding Principles that were developed earlier. These are not mandatory, but experience has shown that it is worth revisiting these to validate them during this task. The reasons for this are twofold. For one, you are working with a broader set of stakeholders. For another, some changes may be found necessary as you get more specific about the processes that will be implemented.

It is recommended that a working session be held to review the Principles. Start first with the Overriding Principles as they were developed in the Visioning Work Stage. Then move to a starting short list. The group can then select, add, delete and modify until the desired set of Principles is drafted.

Any changes to Overriding Principles should be controlled with the Governance Team. This is needed to handle coordination and agreement since many process areas may have changes all at the same time.

Define Inputs and Outputs Task

In this task you will be defining the inputs and outputs of each process. Much of this can be found throughout ITSMLib within each process library. When building these, the suggested approach should be used:

1. Each Process Team defines the key activities for their process

2. Each Process Team defines the inputs and outputs for their process

3. Process Teams then get together in a working session to integrate inputs and outputs among all processes

4. Process Teams then work independently again to refine their inputs and outputs based on feedback from the working session

5. A final workshop is held with all process teams to confirm integration between all the processes

Key Activities for each ITSM process can also be found throughout ITSMLib. You can also draw on much of the existing ITSM literature to expand upon these as needed. Other considerations with this Work Task might include:

- Use of process modeling tools to assist in documenting and integrating the inputs and outputs

- Creation of Use Cases to test inputs and outputs.

A Use Case is situation or scenario that reflects something already experienced in the company that will require the processes you are building to handle. An example of a Use Case might be a high severity incident that is called into the Service Desk. You would then talk through the chain of events that must take place to handle it. Each process team would describe their role in processing the incident and validating the inputs and outputs along the way. This is an excellent way to test what you are designing to ensure it will be fit for purpose.

Define Policies Task

In this task, you will be defining the ITSM policies. A policy is a set of rules or actions that will be adopted by ITSM solutions that will govern how services are to be delivered. The Principles defined earlier can also serve to become policies as well. An example policy might be something like:

> *The Service catalog will be the only authorized source for all Services and Service Level targets for the company.*

It is recommended that a common template be developed for describing all policies. This template might include:

- Policy Statement
- Definition of terms used with the policy or its operation
- Effective date
- Who is effected
- Rationale for the policy
- Background examples as to why the policy needs to be in place
- Example of compliance with the policy
- Example of violation of the policy
- Consequences of violation of the policy
- Contact role for questions about the policy

The approach for developing your ITSM policies should be done as follows:

1. Each process team should review the Overriding Principles and process specific principles developed earlier to determine if they will result in policies that will be needed for their individual processes

2. Each process team independently drafts their own processes

3. Conduct a workshop with all process teams to review each policy that has been developed for buy-in and acceptance across all teams

4. Roll agreed policies up into an ITSM Policy Guide that will cover all processes

Define Procedures Task

Procedures will be developed out of both Key Activities and Policies. A Procedure is a set of Work Tasks and Work Steps to produce an outcome such as producing a Work Product or processing an input. Process activities will consist of one or more procedures that must take place for them to be accomplished.

In this task, you will be developing specific procedures that will be needed for each ITSM process. It is recommended that a common template be used for this activity across all ITSM processes. This template might include:

- Procedure name
- Description
- Work Tasks
- Work Steps
- Roles to carry Out Each Work Step (Roles will be defined later)
- Scope
- Frequency
- Any constraints if they apply
- Dependencies (such as other procedures)
- Procedure inputs and outputs
- Pictorial Workflow

Other considerations for this task include:

- Use of process modeling tools to build and document procedures

- Using Use Cases to test procedures to ensure they will be fit for purpose

- Integrating procedures into an overall process workflow

Define Tools Task

The key output of this task is to define the tooling direction and strategy that will be used to support the ITSM processes. Input to this task will come from the high level architecture developed in the ITSM Visioning Work Stage and the Technology Assessments and Existing Tools Inventories done in the ITSM Assessment Work Stage.

Each process team should review these outputs with the technology teams in light of the process design work done in this Work Stage. From this, identify any tooling gaps that may exist. Then confirm the tooling direction and ensure it is integrated with the overall architecture strategy. It is possible that this later design work might have identified additional requirements not foreseen in the earlier work stages.

The tooling direction should be done individually by each process team first and then reviewed with all process teams. The Technology Track will have a key responsibility herein terms of sorting out and prioritizing all tooling requirements and developing the overall technology support architecture.

Tactical work such as product selections, technology procurement or tool implementations, if they will be needed, will be done as part of the Initial Wins Work Stage.

Define Roles/Skills Task

In this task, needed roles skills and skill levels will be defined that are necessary to carry out ITSM processes and procedures. The ITSM Transformation Organization chapter of this book describes roles in detail. It is in this work task that you will be defining them. That chapter also lists a number of ITSM roles that you may wish to start with and adapt to your particular solution.

It is suggested that roles be developed individually by each process team first. A workshop should be conducted that reviews each role developed. These should then be merged and rolled into a set of ITSM role descriptions that everyone will use.

For each role defined, also list key activities to be performed by that role. This will help in making sure everyone understands what the role will do. It will also serve as input for developing job descriptions which will occur in the ITSM Transition Work Stage. The Organizational Considerations chapter also lists some key activities with the roles presented.

Once roles are defined, the next step is to identify skill traits and skill levels needed for each role. Again, the Organizational Considerations chapter presents some sample skill traits and a simple rating system for defining skill levels. Once defined, these should be confirmed by all process teams.

You may wish to consider adding an additional task to map the roles defined in this task with all the procedures defined earlier. This is a great way to make sure that you have all needed roles identified. It also makes it clear which roles will carry out which ITSM procedures and working tasks.

Define Organization Task

Once roles, skill traits and skill levels have been defined, you can begin the tricky process of mapping these into your future state service organization. You can start by first defining the needed organization structure from the solution point of view. This structure would show the needed hierarchies and reporting structures that the teams feel would best serve the future state solution.

As a next step, you can then look at your current organization structure and see how close it comes to your desired state. If there is a fairly close match, you will be in good shape. If not, you have two choices:

- Propose a different organization structure which will be put into place to support ITSM support and delivery services

- Consider a transitional structure which will be put into place to support ITSM support and delivery services

The first option gets you there faster, but might be fraught with risk if the gap is large between your present organization and the proposed future state. The risk is that the organization is not ready for large changes in reporting structure, roles and responsibilities and chaos might result.

The second option means that an optimum organization structure will not be in place right away, but provides it does provide benefits in easing the organization towards the future state minimizing any chance for chaos. An example of this is to create an ITSM support organization as a virtual team first, then move it to a direct reporting structure at a later date.

Make sure that you also identify any skill gaps that exist. From these, fashion a training strategy and program to address them.

Finally, document whatever decisions you make by developing an Organizational Strategy for your future state. This should consist of:

- The organization structures you plan to adopt

- Transitional activities to get to the desired organization

- Role descriptions

- Skill traits and levels

- Identification of skill gaps

- Training plan to close those gaps

The ITSM Transformation Organization and Transition Work Stage chapters discuss this in more detail and also presents a role mapping approach that you may wish to consider.

Build Process Guide Task

This task simply consists of building the process guides for each ITSM process. The guides themselves will be populated with much of the output from previous tasks in this Work Stage. This may include:

- Process Mission and Goals
- Critical Success Factors and Key Performance Indicators
- Key Activities
- Inputs and Outputs
- Procedures
- Design Principles
- Process Roles
- Role Activities
- Key supporting architecture
- References to Policies in the ITSM Policy Guide

Prior to starting this task, it is recommended that a Process Guide template be developed for documenting all the ITSM processes. Process Guides can also be reviewed by peer Process Teams for accuracy—especially where one process is dependent upon another. ITSMLib provides many example process guides for almost every ITSM process.

Define Process Reporting Task

In this task, design activities will occur to establish reporting on quality metrics for the ITSM processes. These will be used for two purposes: report on quality of processes and services to company management and to seek out continuous service improvement over time.

Critical Success Factors (CSFs) and Key Performance Indicators (KPIs) were developed earlier. This task is concerned with developing the processes, tools and organizational roles and responsibilities for reporting on them. This means identifying:

- Data sources

- Data summarization and calculations that will take place

- Reporting templates

- Tooling support—which tools will be used to produce reports, extract data, provide report and data repositories, etc.

- Procedures for administering the reporting

- Integrating report procedures with other ITSM processes

- Implementation of process reporting solutions will occur in the ITSM Control Work Stage.

Build Transition Strategy Task

This task plays a key role throughout this Work Stage. As stated earlier, the Initial Wins Work Stage is taking place in parallel with this Work Stage. Since Initial Wins are sets of tactical implementations, it is this task that will ensure they are kept in alignment with the strategic process, technology, organization and governance solutions being developed in this Work Stage.

The key objectives of the activities in this Work Task are:

- Ensure Initial Wins are aligned with the strategic solutions being developed for ITSM

- Develop transition plans for both tactical and strategic solutions

Key transition decisions will need to be made for each ITSM solution that will be put into place. That decision is to determine what overall transition strategy will be used. Several options are:

- Direct Cutover
- Phased Cutover
- Parallel Cutover
- Phased Parallel Cutover

These are discussed in further detail in the ITSM Transition Work Stage chapter.

The key output of this task is an Integrated ITSM Transition Plan. This plan will essentially be the key driver for the activities that will take place during the ITSM Transition Work Stage. You may also wish to leverage transition activities both in this Work Stage and the Transition Stage with whatever Release Management plans you are putting into place. This is an excellent way of testing them.

Chapter 10

ITSM Initial Wins Work Stage

"A mountain is made of many small specks of dust. By making small improvements, workers can chip away at the mountains of wasted hours, frustration, and delay".
—Old Japanese Proverb

The main goal of the Initial Wins Work Stage is to implement small short term ITSM related solutions that provide immediate payback and benefits. By doing this, benefits can be achieved early on in the implementation program that will gain positive recognition by the organization and push momentum forward towards the objectives of the ITSM future state vision.

Each ITSM process will typically have several Initial Win projects that came out of the ITSM Assessment Work Stage. Remember that it was in this work stage that Initial Wins were selected, and prioritized.

There are typically is not a large number of these, usually anywhere from 2-5 Initial Win projects per ITSM transition wave. The rule of thumb, as used in the Assessment Work Stage, is to implement those projects that will provide most of the benefits with the least amount of cost and labor. Examples of Initial Win Projects are shown later in this chapter, but here is a short list to give you an idea:

- Building a Service Catalog (Service Catalog Management)

- Holding a contest for best IT cost saving idea (Financial Management)

- Establishing a Problem Analysis Toolkit (Problem Management)

- Implementing standardized Incident Classification Categories (Incident Management)

- Implementing SLAs and OLAs with key business units (Service Level Management)

You can see that the above examples reflect tactical tangible things that can be built, touched and recognized by everyone involved. They also provide benefits that will serve to demonstrate accomplishment and move your organization forward towards your planned future state IT Service Vision.

Initial Win Projects are tightly tied to efforts taking place within the ITSM Planning, Design and Transition Work Stages. That is why the Initial Win Work Stage runs in parallel with these during implementation. Without this, you would risk implementing a set of discrete standalone projects that are not tightly integrated with your overall implementation effort and aligned with your future state IT Service Vision. The Design Stage efforts are used to guide design decisions as you implement each Initial Win. The Transition Work Stage is used to coordinate Initial Win implementation and rollout activities (kind of like Release Packaging!)

Not all Initial Win projects have to always represent a final ITSM solution. It is possible to implement short term solutions that will be expanded on in later years as necessary. For example, you may decide to customize an existing support tool in a manner that supports several service management functions as an Initial Win project early on. Later on, several years down the road, you plan to replace this tool entirely with a tooling suite that more directly contains service management functionality. In an actual use of this, one company implemented Problem Management classification categories to their existing Incident management tool as an Initial Win effort recognizing that the tool would be replaced in two years.

Selected Initial Win projects usually involve all of the Work Tracks: Process, Organization, Technology, Governance and Management. Most can be done in a 2-6 month timeframe and they have a clear sense of accomplishment when completed.

Initial Win Project Lifecycle

Every Initial Win project follows a standard project lifecycle that guides the tasks and activities within it. This lifecycle follows a standard ITSM Release Management process and can be pictured as follows:

Initial Win Project Lifecycle

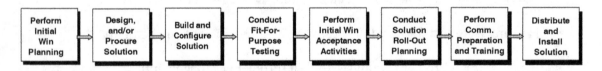

The above lifecycle is used for every Initial Win project identified. This provides structure for each win project and ensures that everyone follows a similar implementation path. This will make communications and status reporting much easier across the teams and ensure that solutions are taking a holistic view into account. Use of the lifecycle also provides an added bonus. It will allow your organization to gain an early hands-on preliminary feel for how parts of the Release Management Process will work.

Each Initial Win Project will appear on a program master schedule that will be managed by the Program Office (a concept discussed in the Organizational Considerations chapter in this book). This will ensure that there is coordination across these projects and that conflicts do not occur.

As can be seen, the activities within the Initial Wins Work Stages are fairly straightforward. They are essentially a series of short term implementation projects that are taking place within each ITSM Process. At the end of each one, however, your organization will have gained some recognized benefits and taken a noticeable step towards the future state ITSM Vision.

ITSM Initial Wins Work Stage Approach

Initial Win projects may be as short as 1 week to as long as 32 weeks within the ITSM transformation effort. The length depends upon the size, scope and complexity of the Initial Win project. Every Initial Win Project, no matter how large or small should follow the sequence of Work Tasks shown below. The timeline shown below is just an example, but should be more accurately determined during the ITSM Planning Work Stage.

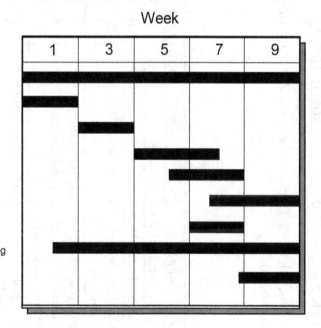

Coordination across all Initial Wins projects should be handled by the ITSM Program Office. A detailed Work Breakdown Structure is depicted as follows:

ITSM Initial Wins Work Stage—Work Breakdown Table

R=Responsible (For Initial Wins, all Work Tracks are shown R but for display purposes—only one track should be responsible depending on the Initial Win chosen)
A=Assists
C=Consults
I-Informed

Work Task	Work Step	Org	Proc	Tech	Gov	Mgt	Work Product
	Manage Initial Win activities	A	A	A	A	R	
	Report Initial Win status	A	A	A	A	R	Status Reports
	Coordinate and schedule Initial Win meetings	A	A	A	A	R	
Manage Initial Win Projects	Integrate Initial Win activities with Planning, Foundation and Control Work Stages as needed	A	A	A	A	R	
	Integrate Initial Win rollout plans with all other Initial Win projects	A	A	A	A	R	
Perform Initial Win Planning	Review Initial Win solution work plan	R	R	R	R	R	
	Modify plan as necessary	R	R	R	R	R	Updated Initial Win Plan
	Integrate plan with Program Office	C	C	C	C	R	Integrated Initial Win Plan
	Design Initial Win Solution	R	R	R	R	I	Solution Designs
	Identify solution roles and responsibilities	R	A	A	A	A	Solution Roles/ Responsibilities
Design and/or Procure Solution	Design needed forms and reports	R	R	R	R	I	Solution Forms and Reports
	Identify solution items to be procured	R	R	R	R	I	Procurement Specifications
	Procure solution items	R	R	R	R	R	Procured Solution CIs
	Perform functionality testing as needed	R	R	R	R	I	Functional Test Results

Work Task	Work Step	Org	Proc	Tech	Gov	Mgt	Work Product
Build and Configure Solution	Implement solution in test environment	R	R	R	R	I	
	Customize solution as needed	R	R	R	R	I	Implemented Solution in Test Mode
Conduct Fit-For-Purpose Testing	Identify test criteria	R	R	R	R	I	Solution Test Criteria
	Conduct test planning	R	R	R	R	I	Test Plans
	Perform solution basic functionality tests	R	R	R	R	I	
	Perform system testing as needed	R	R	R	R	I	System Test Results
	Perform integration tests as needed	R	R	R	R	I	Integration Test Results
	Perform functionality testing as needed	R	R	R	R	I	Functional Test Results
Conduct Initial Win Acceptance Activities	Conduct User Testing of solution	A	R	A	A	I	User Test Results
	Obtain User sign-off on solution	A	R	A	A	I	User Solution Sign-Off
Conduct Solution Rollout Planning	Define solution rollout strategy	R	R	R	R	C	Solution Rollout Strategy
	Identify rollout tasks	R	R	R	R	C	
	Identify rollout roles and responsibilities	R	R	R	R	C	
	Integrate rollout plans with overall solution plans and activities	R	R	R	R	C	
	Document rollout plans	R	R	R	R	C	Documented Rollout Plans
Perform Communication, Preparation and Training	Integrate solution plans with overall implementation communication plans	R	A	A	A	A	Integrated Solution Communication Plan
	Conduct solution training	R	A	A	A	I	Trained Solution Users
	Communicate solution activities and plans	R	A	A	A	A	Solution Communications
	Execute solution rollout activities	R	R	R	R	C	Implemented Solution
Distribute and Install Solution	Provide early solution implementation help and support as necessary	R	R	R	R	I	
	Confirm solution is operable as planned	R	R	R	R	C	

Initial Wins Key Work Products

Work Products will vary greatly depending on the solution implemented. Work Products for each solution should be identified during the solution planning stages in the ITSM Assessment Work Stage. The table below presents generic Work Products that solutions should typically deliver:

Work Product	Description
Integrated Initial Win Plan	Work plan that shows tasks, timeframe and responsibilities for implementing the Initial Win Identifies needed skills and training to execute solution Integrated with overall Program Office plans and activities
Solution Designs	Documented solution plan Includes design charts, workflows, forms, reports as needed Indicates planned solution support technologies as needed Identifies solution related roles and responsibilities
Solution Test Criteria	Identifies solution acceptance criteria Indicates planned test results
Test Plans	Documented plans to test solution Identifies roles and responsibilities for testing Indicates testing schedules and timeframes Identifies planned test runs Might include capacity, performance and functionality tests
User Solution Sign-Off	Represents formal acceptance of solution Should be formally documented and signed off by one or more business unit representatives
Solution Rollout Strategy	Identifies strategy for how solution will be rolled out Should be integrated with all other Initial Wins and implementation activities Identifies roles and responsibilities for rollout activities Might include an early launch strategy (i.e. pilot)
Solution Communications	Communicates solutions and activities to those impacted by them Activities should be integrated with overall ITSM implementation Communication Plan and activities Should contribute towards user acceptance of solution

In addition to the above, each Initial Win may have unique Work Products that are relevant to it such as a Report, Implemented Policy or software tool.

ITSM Initial Wins Implementation Team

The team organization for each Initial Win will be unique depending on the type of solution being implemented. It is recommended that virtual teams co-exist inside the overall ITSM Transformation Organization to implement Initial Wins. These teams should be process aligned. By this, for example, all Initial Wins that belong to a process are guided, managed and controlled by the relevant Process Owner.

The following chart shows a recommended Initial Win Virtual Team Organization:

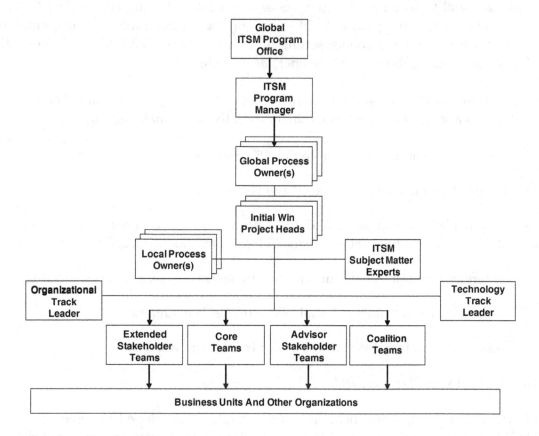

The Virtual Team organization closely represents the overall ITSM Transformation Organization (See Organizational Considerations chapter). The one exception is the addition of an Initial Win Project Head position as shown in the above structure. This position serves as a lead and coordination point for one or more Initial Win projects and works directly with the Process Owners. Each Process Owner would oversee all Initial Wins within its scope. The Program Office would coordinate Initial Win efforts for all ITSM processes.

Note that the above organization is rather large and may not be necessary for Initial Wins that are smaller in scope. Feel free to use a subset of this as needed.

ITSM Initial Wins Work Task Considerations

This section describes each of the ITSM Initial Win Work Tasks in a little more detail. You can also consult existing ITSM documentation sources for Release Management that exists in the market place. Much of the general tasks described in this literature also apply to implementing Initial Wins.

Manage Initial Wins Work Stage Task

This task will focus on overall management of all Initial Win projects. The Program Office will do all the typical tasks of setting up the project, time reporting, obtaining work space and meeting rooms, setting up status reporting and other administrative functions to ensure this Work Stage operates smoothly.

Initial Win projects will start and complete all at different points of time. The Program Office will also need to coordinate Initial Win activities to make sure that:

- There are no conflicts between Initial Win projects

- Project dependencies are managed and addressed

- Stakeholders do not get asked to perform tasks in a chaotic fashion (i.e. one stakeholder does not get daily requests for information from different projects)

- Synergies between Initial Win activities are leveraged (i.e. a common meeting)

- A consolidated schedule of Initial Win activities is maintained

- Issues between Initial Win teams are addressed and resolved

Perform Initial Win Planning Task

Detailed Initial Win implementation plans should be developed in conjunction with ITSM Planning Work Stage activities. Work Plans should detail implementation tasks, roles, responsibilities, key milestones, work products and timelines for Initial Win activities.

Plans should be developed for each Initial Win activity separately. These should then be rolled up into an overall plan for each ITSM process. Another rollup should then occur to roll all Initial Win projects for every process into the Program Master Plan.

Design and/or Procure Solution Task

In this task, design activities should take place for each Initial Win solution. Be sure to consider your designs in a holistic fashion. By this, make sure you are addressing Process, Organization, Technology and Governance for each solution being developed.

Make sure designs are reviewed appropriately by key stakeholders that will be impacted by the solution. This means reviewing solution designs with Subject Matter Experts, business unit representatives and key decision makers. This will serve to enhance acceptance of your solutions.

Procurement activities represent a potential risk to ITSM transition timelines and schedules. Make sure you determine what needs to be done in this area as early as possible to avoid delays.

Build and Configure Solution Task

In this task, the planned Initial Win solutions will be implemented in a test manner. For technology solutions, this means building and configuring them within a test environment. For process solutions, this might mean establishing a process with a test group or within the implementation team.

Seek out synergies wherever possible when building and configuring solutions. For example, a technology solution may be implemented and integrated with other solutions being put into place. Likewise, a process solution may be joined with another process or policy effort.

Conduct Fit-For-Purpose Testing Task

In this task, the solutions are tested. Typically there are four levels of testing:

Basic Functionality Testing

> This level of testing ensures that solutions will perform as advertised and expected when initially implemented. For example, a reporting tool solution will extract a small subset of test data and produce a basic report. In many cases, vendor solutions may come with basic functionality tests to ensure they are operating correctly.

System Testing

> This level of testing ensures that the solution operates correctly from a technology point of view after all customization functions have been put into place. Using the reporting example from before, this would mean making sure that desired data is extracted and formatted into planned reporting templates and that reports are correctly displayed and printed.

Integration Testing

> This level of testing ensures that the solution operates correctly when put together with other system infrastructures and components. In our reporting example, this might mean conducting tests to make sure that reporting tools

work with needed databases and external data repositories and whether report output can be used as input to other systems in the company.

Functionality Testing

This level of testing ensures that the solution functions correctly and meets the needs of the business. In our reporting example, this might mean conducting tests with end users to make sure report totals have been calculated correctly and that useful report metrics are being reported on.

Testing of processes and policies can be done by using Use Cases within the project team first. This can be done by having a working session with key team members and stakeholders. Each participant is assigned a role in that session. Use Cases are presented one at a time and the participants play act their role in handling it. This is an excellent way to test processes and policies for usefulness and missing actions or roles.

Once Use Case testing is complete and participants are satisfied with the results, the process/policy can be issued as an Early Launch (Pilot) effort with a selected business unit or IT department. The solution is then monitored by the Project Team and any deficiencies are addressed. Once results are where the team wants them, the solution can be considered ready for deployment across the organization to all intended stakeholders. Early Launch efforts should be leveraged with activities in the ITSM Transition Work Stage.

Conduct Initial Win Acceptance Activities Task

In this task, activities are carried out to have key stakeholders and approvers agree the solution operability. A formal acceptance procedure and sign-off should take place for this. The key thing is to make sure there no misunderstandings and misconceptions about what is going into production.

This is typically done twice for each solution. The first time indicates an acceptance that the solution is ready for deployment to business units and IT departments. The second time is to validate that the solution is reasonably acceptable after deployment activities have completed.

Conduct Solution Rollout Planning Task

In this task, rollout and deployment plans are developed and published to the Program Office. Rollout plans are somewhat like implementation work plans. They identify activities, roles, responsibilities and timelines for deployment and transition activities. Rollout plans should be holistic in nature and include all aspects of Process, Technology, Organization and Governance in their execution.

As before with the work plans, rollout plans should be rolled up to the process level first and then integrated into the overall Program Office master Plan.

Key considerations for rollout planning might include:

- Distributing software to servers and workstations

- Training users on new technologies and processes

- Training IT support staff on new technologies and processes

- Communication activities

- Timelines for rollout activities

- How rollout progress reporting will be performed

- Considerations for site survey and preparation before a solution is implemented

- Ensuring local procurement activities are being addressed on a timely basis

- Production of guides and instructional information to business units and IT staff

- Transition strategy (discussed in ITSM Foundation chapter)

- Roles and responsibilities for rollout activities

- Security issues and considerations during rollout activities

- Additional support needs and handholding during transition activities

Perform Communications, Preparation and Training Task

In this task, communications and training activities will take place around the Initial Wins being put into place. These should occur in accordance with overall ITSM communication plans and activities. Some key considerations might include:

- Ensuring IT support staff is adequately trained to support solutions

- Ensuring users of solutions being developed are adequately trained in new processes and technologies if these apply

- Ensuring users of solutions are aware of when solutions are going into place and what the impacts will be

- Ensuring users of solutions are aware of any activities they must undertake prior and after solutions have been put into place

- Ensuring users of solutions are aware of when solutions are being put into place

- Providing adequate feedback mechanisms for user issues and complaints

- Keeping Senior Management aware of solution deployment progress and any issues that may exist

Distribute and Install Solution Task

In this task, the Rollout Plan activities are executed. The project team and ITSM Program Office should oversee and monitor these activities to make sure they are being performed as planned. Plans and timelines should be updated as needed throughout the rollout effort in the event delays or changes occur.

Initial Win Examples

Overview

This section briefly lists examples of Initial Win Projects for some ITSM processes. These are provided as a starting point for developing your own.

Incident
Management

Implement an Incident Classification Schema
Implement an Incident Escalation Policy
Implement an Incident Alerting Strategy
Implement Incident Logging and Recording Audit Procedures
Implement an Incident Management Knowledge Base
Implement an Incident Logging Tool
Generate logged incidents from system events

Problem
Management

Conduct Problem Management Awareness and Education Campaign
Implement Problem Classification Schema
Implement Problem Trend Reporting
Implement Root Cause Analysis Workshop
Assemble Problem Analysis Techniques Toolkit
Implement Standard Problem Resolution Activities Lifecycle
Implement Problem Escalation and Communications Policy

Change
Management

Implement a CAB
Implement a Change Management Quality Assurance Checklist
Build and Document a Change Policy
Implement a Change Impact Analysis Procedure
Implement a Forward Schedule of Changes
Create and Publish a standard Request For Change (RFC) Form
Implement an Emergency Change Approval Procedure
Implement a Change Communications and Escalation Policy
Implement a Change Tracking Tool

Release
Management

Build and implement a Release Policy
Build a Definitive Software Library (DSL)
Build a secure Definitive Hardware Store (DHS)
Implement a Release Schedule for all known releases
Implement an Operational Acceptance Checklist
Implement a Release Management Awareness Campaign
Implement a Patch Management Policy

Configuration Management	Define a Configuration Management Database (CMDB) Schema Model
	Implement a CMDB Infrastructure Solution
	Identify, classify and store all known IT documents of reference
	Identify all known configuration repositories and provide links via website
	Document CI types and relationships for key IT services and solutions
	Implement auto-discovery tool
Service Level Management	Establish Service Level Agreements
	Establish Operational Level Agreements
	Obtain and store all official copies of Underpinning Contracts
	Establish Business Unit Service Representation
	Implement Service Reporting and Review procedures
	Implement reporting on process metrics
	Implement a Service Management Dashboard
	Implement a Service Escalation Policy
	Establish and staff a Service Manager organization position
	Establish and staff Process Owner responsibility positions
	Establish single point of contact role for each IT operational unit
	Implement a Service Improvement Program (SIP)
Availability Management	Establish Availability SWAT Team to address key availability issues
	Establish and staff a single point of contact role for service availability
	Implement a Service Risk Analysis procedure
	Establish Availability Guidelines for IT solutions
	Conduct a Service Risk Analysis for new or existing IT solution
	Establish an Availability Analysis Techniques Toolkit
	Implement availability monitoring solution for key services
	Define what "availability" is and how to calculate it for the enterprise
	Implement an availability solution for a high risk service
Capacity Management	Implement a Capacity Modeling Tool
	Implement Workload Characterization for key IT Workloads
	Identify business drivers that impact capacity for each service
	Establish periodic Capacity Reporting
	Publish Performance and Capacity Guidelines for new IT solutions
	Document and publish capacity assumptions and estimates
	Implement a Capacity and Performance Review procedure
	Implement a capacity simulation tool for IT test environments

IT Service Continuity Management	Establish an IT Continuity Plan Establish an IT Continuity Strategy Establish an IT Continuity Awareness and Education Campaign Conduct mock disaster and identify risks found Establish IT Service Continuity representation in CAB meetings Establish Underpinning Contract with IT Service Continuity provider
IT Financial Management	Build an IT Cost Model Implement charging for IT services Establish a contest for best cost cutting solution with IT staff Implement financial reporting integrated with corporate ledger Identify and inventory IT cost drivers Identify and eliminate IT assets and licenses not in use Define and publish an IT Charging Policy
Service Desk	Build a Service Desk strategy (i.e. Virtual, Local, Central Desk) Implement a Call Management solution Implement a Staff Coverage Estimating Model Define and Publish a Service Desk Escalation Policy Document Service Desk Roles, Skills and Job Descriptions Implement Customer Satisfaction Survey Implement Service Desk Staff Training Program Conduct a Customer Relationship Skills Education and Awareness Program
Security Management	Implement a Security Policy Implement a Security Monitoring Solution Conduct a Security Risk Assessment and Analysis Integrate disparate security functions into virtual security services Resolve selected existing security audit issues Conduct Security Awareness and Education Campaign
Service Catalog Management	Build a service catalog Create IT service descriptions for consumers Market IT services

Initial Win Description Example

| Overview | This section presents a template for briefly describing a sample Initial Win. |

See following pages:

Attribute	Definition
Project	Establish, Build and Implement an IT Service Catalog
Immediate Benefits	Clear understanding of services needed to support the business Clear understanding of business needs and requirements Clear understanding of capabilities needed to support the business
Work Breakdown	Identify Services that delivers to its customers Identify and agree Service Catalog structure and schema Identify and implement tool support strategy for Service Catalog Populate Service Catalog with known service information, sub-services and customers Identify candidate service targets for each catalog service Define Service Catalog maintenance procedures Conduct Service Catalog Awareness Campaign
Work Products	Inventory of services, service targets and customers Service Catalog tool support strategy Populated Service Catalog Service Catalog maintenance procedures Service Catalog Awareness Campaign
Assumptions	Main focus will only be on services that need to be managed versus all services Sustained commitment to formalization of roles and responsibilities Sustained management support for accountability Participation with needed training and awareness efforts
Risks	Acceptance of roles and responsibilities Ensuring awareness activities cover all needed parties impacted by Service Catalog usage and maintenance Ensuring agreement on Service Catalog information
Success Metrics	Number of services in Service Catalog Number of personnel trained in Service Catalog usage Number of RFCs raised with Service Catalog changes
Duration	Approximately 8 weeks

Estimated Staffing	Skill Set/Role	Participation	Total Days
	Project Manager	Part Time—½ day per week	4
	Operations Architect	Full-Time—5 days per week	40
	Subject Matter Expert	Part-Time—2 days per week	16
	Organization Change Analyst	Part-Time—2 days per week	16
	Technical Writer	Part-Time—2 days per week	16
	Stakeholders (Extended Team)	Part-Time—½ day per week	4 each

Chapter 11

ITSM Transition Work Stage

> "Trust is good, but control is better."
> —Vladimir Lenin

A bad measure can cause bad decisions . . . .

The goal of the ITSM Transition Work Stage is to implement the ITSM solutions designed in the ITSM Design and Initial Wins Work Stages and begin the regular cycle of control reporting on their health and state. This is the Work Stage where the pedal hits the metal in terms of moving designed processes outside the implementation teams to an active state and implementing the necessary cultural and organizational changes necessary to support desired ITSM activities.

Efforts in this Work Stage will focus around transition activities for your ITSM solutions. Up to this point, solutions have been designed and tested. It is in this Work Stage that they become deployed, operated and reported on. It serves to guide and control how those items are implemented as part of a holistic set of ITSM solutions and integrated with the rest of the organization.

Implementing processes is not the same as implementing software or hardware. A process is not considered implemented unless people are actually using it. Process metrics and reports will provide evidence that this has actually occurred. Therefore, some of the key activities in this Work Stage will include:

- Executing ITSM solution Transition Plans

- Training personnel on how to use the new processes and solutions

- Implementing and reporting on the process metrics

- Implementing new job descriptions

- Implementing organizational changes

- Assigning personnel to ITSM job roles

- Activating ITSM Governance activities such as service reviews

At the end of this effort, all Initial Win projects will have been completed and the IT organization will be operating with ITSM processes, technologies and organization responsibilities. The ongoing cycle of metrics reporting, service reviews and service improvement initiatives will have been started.

All this leaves two final questions that need to be answered as a last step:

- Has the organization achieved the Implementation Program goals set out at the beginning of the effort?

- Which ITSM service enhancements should the organization focus on for next year?

Answering these questions will involve taking a step back and reviewing the transformation program metrics developed in the ITSM Vision Work Stage. Have program benefits for this year been achieved? Finally, an inventory of action items to take next year is developed. Which of these should the organization plan to implement in the coming year?

Completion of the ITSM Transition Work Stage marks the end of first ITSM wave effort. As part of an ongoing improvement cycle, you can circle back to the Vision and Assessment Work Stages to identify and plan for those activities to be done in the next wave. From this, you can determine additional Initial Win activities for the following wave. This process can be repeated to continually examine your progress towards the future state ITSM Vision year after year.

ITSM Control Work Stage Approach

If we look at the Transition effort over an approximate 16-17 week timeline, the key Work Tasks might be timed to look like the following:

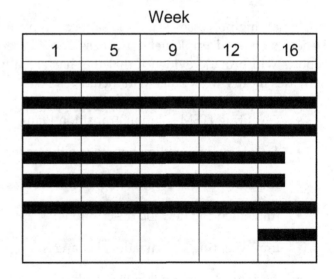

A detailed Work Breakdown Structure is shown as follows:

ITSM Transition Work Stage—Work Breakdown Table

R=Responsible
A=Assists
C=Consults
I-Informed

Work Task	Work Step	Org	Proc	Tech	Gov	Mgt	Work Product
Manage Control Work Stage	Manage Control Stage activities	A	A	A	A	R	
	Report Control status	A	A	A	A	R	Status Reports
	Coordinate and schedule Control meetings	A	A	A	A	R	
	Integrate Control activities with Initial Wins as needed	A	A	A	A	R	
	Integrate Control activities across all processes	A	A	A	A	R	
	Identify Initial Win synergies as needed	A	A	A	A	R	
	Finalize transition solution packages	A	R	A	A	A	Master ITSM Rollout Schedule
Coordinate Initial Win Rollout Activities	Execute Initial Win transition activities	A	A	A	R	C	
	Integrate ITSM solution with existing technology infrastructure						
	Integrate ITSM solutions with organization	R	A	A	A	C	
	Obtain ITSM solution approvals and signoffs	A	A	A	R	C	Implemented and Agreed ITSM Solutions

Work Task	Work Step	Org	Proc	Tech	Gov	Mgt	Work Product
Coordinate Communications and Training	Execute ITSM Training Plans	R	A	A	A	I	Trained IT Staff and Users
	Coordinate Communication Plan Activities	R	A	A	A	C	Executed Communication Plan
	Identify and resolve stakeholder issues	R	C	C	C	C	Resolved Stakeholder Issues
Implement Service Reporting	Establish Service Reporting	C	A	A	R	I	Service Reports
	Establish Process Reporting	C	A	A	R	I	Process Quality Reports
	Establish and Consolidate Operational Reporting	C	A	A	R	I	Operational Reports
	Test and validate reporting procedures	C	A	A	R	I	Implemented Reporting Procedures
	Establish ITSM Management Dashboard	A	A	A	R	I	ITSM Management Dashboard
Implement ITSM Organization	Develop ITSM Job Descriptions	R	C	C	C	C	ITSM Job Descriptions
	Assist with transfers to new organization	R	C	C	C	C	Company Transfers for ITSM
	Assist with recruiting efforts as needed	R	A	A	A	C	New ITSM Company Hires
	Execute Service Organization Strategy	R	C	C	C	C	ITSM Service Organization
Implement ITSM Governance Controls	Establish ITSM Governance Process	A	C	C	R	C	
	Identify governance roles and responsibilities	C	A	C	R	C	Governance Roles and Responsibilities
	Document ITSM Governance Process	C	C	C	R	I	ITSM Governance Process

Work Task	Work Step	Org	Proc	Tech	Gov	Mgt	Work Product
	Establish current Program Metrics results	A	A	A	R	C	Current Program Metrics
	Compare current metrics against baseline	C	C	C	R	I	
Review First Year ITSM Accomplishments	Identify candidate Initial Wins for following year	A	A	A	R	C	Candidate Initial Wins
	Prepare management ITSM Results presentation	A	A	A	R	C	ITSM Program Results Presentation
	Obtain management and stakeholder feedback on ITSM accomplishments and progress	C	C	C	R	C	ITSM Feedback

Transition Key Work Products

The table below presents each Work Product produced by the ITSM Transition Work Stage that is a milestone and provides a further description of it. Other Work Products are described further in the ITSM Transition Work Task Considerations section of this chapter.

ITSM Transition Key Work Products

Work Product	Description
Master ITSM Rollout Schedule	Lists all ITSM Initial Wins and activities Identifies start/stop timeframes for each activity
Integrated and Agreed ITSM Solutions	Implemented Initial Wins and ITSM solutions Solutions integrated with existing process, technology, organization and governance company infrastructures Formal signoff on implementations by key decision makers
Trained IT Staff and Users	Successful knowledge transfer of needed skills to ITSM staff Might include tests or certification to prove knowledge transfer has taken place
Service Reports	Identifies how well SLAs have been met Organized by service or service agreement (SLA) Includes comparison of service target to actual service achieved Documents reasons for large discrepancies from targets
Process Quality Reports	Identifies state and health of the implemented ITSM processes Lists Critical Success factors (CSFs) and Key Performance Indicators (KPIs) and their results Documents process issues Recommends candidate process improvements
Operational Reports	Identifies how well OLAs and UCs have been met Organized by underpinning provider Includes comparison of service target to actual service achieved Documents reasons for large discrepancies from targets Identifies discrepancy impact on service levels May include service improvement plans on behalf of a provider
Implemented Reporting Procedures	Documents workflows for producing control reports Indicates roles and responsibilities for executing reports Integrates with ITSM Governance Process
ITSM Job Descriptions	Describes an organizational position within the company May include part, one or more ITSM roles Describes job activities which can be taken from ITSM role descriptions Should be integrated with corporate Personnel and Human Resource functions

Work Product	Description
ITSM Service Organization	Implies that a set of jobs responsible for carrying out ITSM activities has been established
	Includes organization structure and reporting responsibilities
	Includes people responsible for ITSM Process ownership
ITSM Governance Process	Manages and controls changes to ITSM standards and services
	Describes workflows, roles, responsibilities and tools to be used to support the process
	Includes a research function that looks for new technologies and practices to support ITSM activities
	Identifies an escalation process for resolving ITSM conflicts in scope and architecture
	Should also include process and service reporting functions
Candidate Initial Wins	Identifies further Initial Wins discovered during the implementation effort
	Describes suggested activities that should be done for the following year
ITSM Program Results Presentation	Reports on ITSM Program Success Factors
	Includes Candidate Initial Wins suggested for following year
	Includes lessons learned from transformation activities
	Trumpets all accomplishments achieved as a result of the transformation effort

ITSM Transition Implementation Team Organization

The Team Organization is the same as for the ITSM Design Work Stage. It is pictured again below:

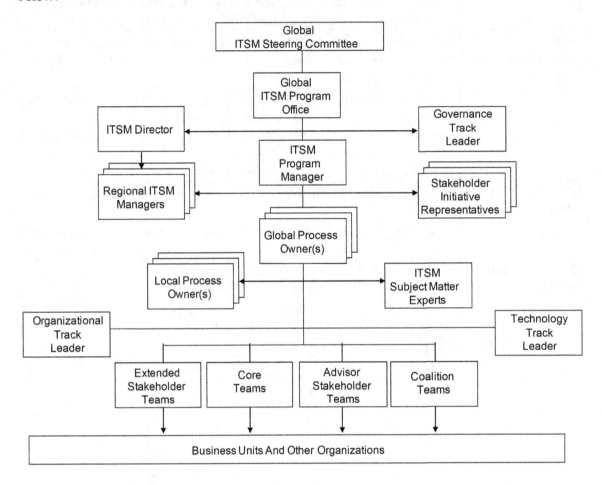

The above Program Implementation Organization structure is discussed in length in the ITSM Transformation Organization Chapter earlier in this book.

ITSM Transition Work Task Considerations

This section describes each of the ITSM Transition Work Tasks in a little more detail. Practical tips and ideas are shown here for each Work Task that you may find useful. Feel free to steal any of the ideas presented here or modify them to suit your specific needs.

Manage Transition Work Stage Task

This task will focus on overall management of the transition effort. As before, the Project Manager will do all the typical tasks of setting up the project, time reporting, obtaining work space and meeting rooms, setting up status reporting and other administrative functions to ensure this phase operates smoothly. Most importantly, this Work Task will ensure that the Transition Team is appropriately staffed, outside consulting resources are procured, if used, and that people are ready to work on it.

Coordinate Initial Win Rollout Activities Task

Since there will be many Initial Win projects and process implementation activities all going on simultaneously, there needs to be an overall coordination effort to prevent chaos. This is the key responsibility of the Program Office. The Program Office should review the plans and activities of each effort taking place and set an overall master schedule.

Synergies between tasks should be looked for to reduce the number of independent contacts with stakeholders. See the Transition Considerations section of this chapter for more detail. Without this, stakeholders may experience daily interactions with different process solutions which can create frustration. The whole implementation effort might appear as if one hand is not sure what the other is doing.

Once this schedule is in place, the Program Office will then coordinate ITSM solution activities throughout this Work Stage. Solutions will need to be integrated with existing company activities around Process, Technology, Organization and Governance.

The following illustrates an example of an ITSM Rollout Master Schedule:

Example Master Rollout Schedule (Sample Page)

Process	Initial Win	1	2	3	4	5	6	7	8	9	10	11	12	Milestone
Incident Management	Implement an Incident Classification Schema	X	X											Incident Classification Schema
	Implement an Incident Escalation Policy		X	X										Incident Escalation Policy
	Implement an Incident Alerting Strategy				X			X	X					Incident Alerting Strategy
	Implement Incident Logging and Recording Audit Procedures											X	X		Incident Audit Procedure
Problem Management	Conduct Problem Management Awareness and Education Campaign				X	X								PM Workshops
	Implement Problem Classification Schema		X											PM Classification Schema
	Implement Problem Trend Reporting				X		X							Problem Trend Reports
	Implement Root Cause Analysis Workshop					X					X	X		Root Cause Workshops
	Assemble Problem Analysis Techniques Toolkit													PM Analysis Toolkit
Change Management	Implement a Change Impact Analysis Procedure									X	X	X		Change Impact Analysis Procedure
	Implement a Forward Schedule of Changes			X										Forward Schedule Of Change
	Create and Publish a standard Request For Change (RFC) Form													RFC Form Template
	Implement an Emergency Change Approval Procedure					X	X							Emergency Change Procedure
Release Management	Build and implement a Release Policy	X								X				Release Policy
	Build a Definitive Software Library (DSL)							X	X	X	X	X		Definitive Software Library (DSL)
Configuration Management	Define a Configuration Management Database (CMDB) Schema Model		X	X	X									CMDB Schema
	Implement a CMDB Infrastructure Solution				X	X	X	X	X	X	X			CMDB Solution
Service Level Management	Build an IT Service Catalog		X	X	X									Service Catalog
	Establish Service Level Agreements						X					X		SLAs
	Establish Operational Level Agreements						X							OLAs
	Obtain and store all official copies of Underpinning Contracts				X									Underpinning Contracts Inventory
	Establish Business Unit Service Representation					X								Identified Business Representatives
	Implement Service Reporting and Review procedures									X	X	X		Service Review Procedure
	Implement reporting on process metrics										X			Process Metrics
	Implement a Service Management Dashboard												X	Service Management dashboard
	Implement a Service Escalation Policy		X											Service Escalation Policy
	Establish and staff a Service Manager organization position							X						Installed Service Manager

The Rollout Master Schedule should be owned and maintained by the Program Office. Changes in timeframes and status need to be communicated on a weekly basis to ensure this schedule is up to date.

This task ends when each item on the Rollout Master Schedule has been transitioned and signed off on by key decision makers.

Coordinate Communications and Training Task

In this task, the ITSM Communication Plan activities are carried out. This includes everything from Program announcements to delivery of whatever training and skills transfer activities were planned for in previous Work Stages.

It is critical at this stage to stay on top of your stakeholders at all times. You need to monitor their issues and satisfaction levels with implementation and transition activities as they progress. The Organization Work Track should act as a focal point for this. Use this track as a contact point for stakeholders to direct issues and complaints to. The Organization Track in turn, should work with the other Work Track teams to resolve issues and communicate status back to stakeholders to the greatest extent possible.

Implement Service Reporting Task

They only way to surely tell whether a process has been implemented and is operating successfully is through metrics. Throughout previous Work Stages, metrics were developed for each ITSM process. It is this Work Stage that reporting on those metrics will be implemented. This means putting into place all the reports, reporting procedures, roles and responsibilities around reporting.

Different kind of reports may be produced depending on service management role or organization unit. These might include reports that show:

- Ability to meet SLA Targets

- Ability to meet OLA targets

- Contractor ability to meet targets in Underpinning Contracts

- Process state and health

In many cases, Senior Management typically likes to see a dashboard report to quickly identify the big picture. An example of one of these is shown as follows:

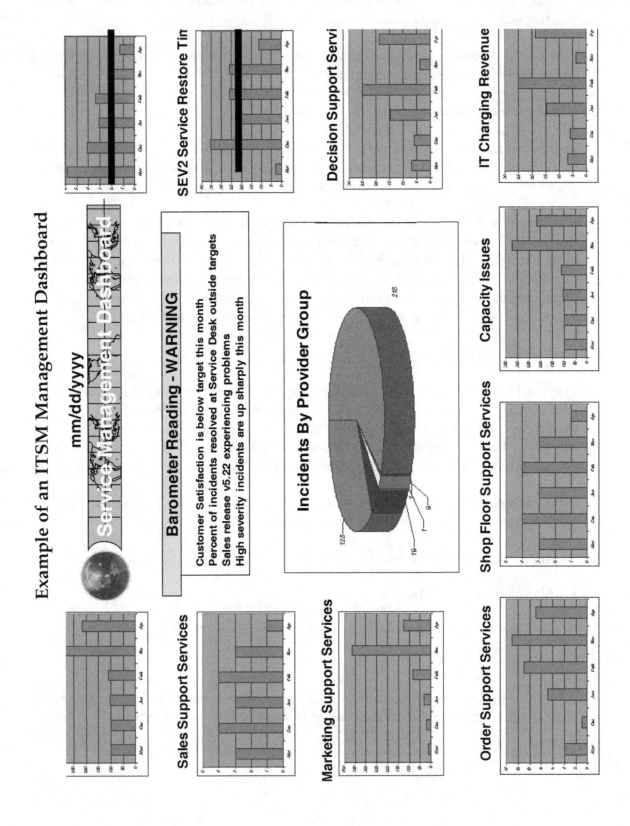

Metrics chosen for dashboard and other management reporting should typically be based on whether decisions and actions need to be taken based on the metric values. Too often, many IT reports to management list historical information that really doesn't tell management whether decisions or actions should be taken.

Reporting solutions chosen should have the ability to be performed with the minimal amount of additional labor and overhead as possible. Considerations should include:

- Avoiding need for dedicated full-time labor to produce the reports

- Avoiding performance or capacity related service disruptions while reports are being generated

- Providing drill-down links if management wants to see more detail

- Being sensitive to the audience that will receive the reports

- Ensuring that reporting is treated as a service just like other IT services

Implement ITSM Organization Task

Prior to this Work Stage, much work was spent by the Organization Track in defining roles and responsibilities for ITSM support services. For this task, those items will now become real by creating job descriptions, job positions, assigning ITSM tasks, obtaining personnel resources and carrying out the overall ITSM Organization Strategy developed earlier.

Note that a job description is totally independent from an ITSM role. Job descriptions might:

- Have one role for the entire job

- Have many roles that are combined in one job

- Have part of a role that is performed by the job

For this task, the Organization Track will typically work with company Human Resources and Personnel departments. This involves developing the job descriptions, making them part of the company's permanent records and recruiting staff as necessary.

Depending on the Organization Strategy, other actions may need to be taken such implementing a service bonus program, changing organization structures, implementing a service quality certification program or transferring resources to different parts of the organization.

Implement ITSM Governance Controls Task

As implementation work proceeds and over time, there will be many challenges to the ITSM solutions being put into place. Business Units and IT departments may start asking for scope changes or additions for new technologies. Changes in business needs may drive further alterations to solutions. An effective way needs to be established to process and approve changes to the ITSM standards and solutions that are being put into place.

Establishment of ITSM Governance process is critical towards managing and controlling services once over time. It is also used to protect whatever ITSM standards and service operations that are being put into place. It makes sure that all appropriate advice, review and consensus have been used to determine whether changes to solutions are valid and necessary.

This is not a substitute for a Change Management process. You can think of a Governance process as kind of a sophisticated escalation process for approving major changes in scope and delivery of ITSM services. In addition, the Governance Process will stay on after the implementation effort is done to handle process reporting, service reviews and the ongoing cycle of service improvement.

A detailed ITSM Governance process is described later in this chapter.

Review First Year ITSM Accomplishments Task

As a final task, ITSM Implementation Program Success Factors should be reported on and reviewed. Remember that these were first developed early on in the ITSM Visioning Work Stage. At this point, you should report on how well those factors were met.

A presentation can be developed to trumpet the successes and lessons learned during the implementation. A section of this presentation should also be developed to describe any candidate Initial Wins for the next wave.

ITSM Solution Transition Strategy Considerations

At the start of the Transition Work Stage, the ITSM Implementation effort is dealing with many Initial Win and Process implementation solutions all occurring at the same time. Chaos would reign if each of these efforts were to be individually transitioned. Stakeholders would experience wave after wave of new changes which could lead to a halt in the entire effort.

For this reason, Initial Wins and process solutions should be grouped and packaged in a manner not unlike what happens in the Release Management process. For example, you may wish to group all ITSM Service Support solutions together as one integrated solution. This integrated solution would contain all the process, technology, organization and governance changes as part of one integrated solution package.

Once these have been developed, the next step is to determine whether solutions are to be deployed across the entire organization simultaneously or via an Early Launch effort. The Early Launch effort, which is highly recommended, involves deployment to a small subset of your business organization first. Once things appear to be working successfully, deployment then goes out to the rest of the enterprise. There are many advantages to conducting an Early Launch effort. Some of these are:

- Opportunity to resolve issues and problems with a much smaller group of people

- Greatly reduces risk to overall services and operations

- Provides benefit testimonials by others not originally associated with the implementation effort

Another key transition decision is how to effect the transition itself. Should it be implemented all at once or in pieces? Several ways to do this exist. They are:

- Direct Cutover

- Phased Cutover

- Parallel Cutover

- Phased Parallel Cutover

Let's take a look at each of these options in a little further detail on the following pages. The charts shown illustrate at a high level how each strategy works.

Direct Cutover Strategy

Direct Cutover

Time Period #1   Time Period #2

| Solution #1 |

| Solution #2 |

Test | Solution #3 |

Production | Solution #1 |

| Solution #2 |

| Solution #3 |

This approach transitions all the ITSM solutions directly at once from where they are being tested right into production. Most organizations like this approach because it is the most direct and fastest to use. However it has some risks.

Pros

- Fastest way to implement
- Cheapest way to implement

Cons

- Requires very solid testing
- No fallback if things go wrong
- Riskiest of all approaches

Good For

- Implementations that are smaller in scope, minimal transition impact on day-to-day services and that do not have large impacts on people and their current day-to-day activities.

Phased Cutover Strategy

Phased Cutover

	Time Period #1	Time Period #2	Time Period #3, etc.

Test
- Solution #1
- Solution #2
- Solution #3 | Solution #3

Production
- Solution #1 | Solution #1
- Solution #2 | Solution #2
- | Solution #3

In this approach, ITSM solutions are cut over individually. In the above example, two solutions have been cut over first. The third solution was not cut over until a later period.

Pros

- Less risk when deploying ITSM solutions

Cons

- Longer time to deploy solutions
- Temporary mechanisms and throwaway solutions might be needed to integrate dependencies between solutions in production and those not yet in production

Good For

- Large implementations that are foreseen to take a long period of time, but where there is a need to show results quickly.

Parallel Cutover Strategy

Parallel Cutover

	Time Period #1	Time Period #2	Time Period #3, etc.
Test	Solution #1	Solution #1	
	Solution #2	Solution #2	
	Solution #3	Solution #3	
Production		Solution #1	Solution #1
		Solution #2	Solution #2
		Solution #3	Solution #3

With this approach, the entire solution set is placed in both test and production at the same time. Operations perform services in two modes: the current way and the future way both simultaneously. This continues until it is discerned that future way is working. At that point, the old way is eliminated.

Pros

- Fallback is in place if there are problems with the new solutions

Cons

- Operationally difficult—staff must do things twice in different ways
- Fixing and adjusting pieces of the future state solution will be hard while in parallel mode
- May have to deal with staff dissent over doing things in parallel
- Requires the most staffing and technology resources of all the transition approaches

Good For

- The situation where solutions must be implemented quickly, but risks require that a fallback solution still be in place to minimize service disruptions to the business.

Phased Parallel Cutover Strategy

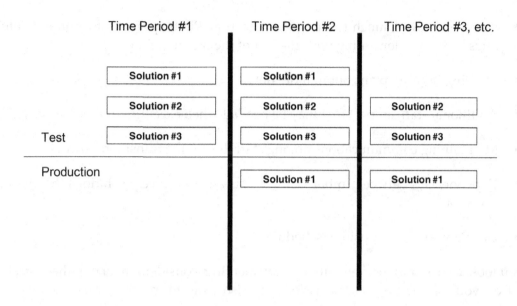

With this approach, solutions are cutover in phases as previously shown, but the old way and future way are only done for those phases that have been cut over.

Pros

- Safest way to transition
- Minimizes the number of things that have to be done both the old way and the new way

Cons

- Most expensive way to effect the transition
- Will take the longest transition time of all the options
- Temporary mechanisms and throwaway solutions might be needed to integrate dependencies between solutions in production and those not yet in production

Good For

- This strategy is best for the situation where the IT service is critical to the business and high levels of service are required. This approach can be used to minimize the risk of the transition and ensure that services are not disrupted while transition activities are taking place.

Other Transitional Considerations

Other considerations for transition of solutions should consider items such as:

- Using an Early Launch (sometimes known as Pilot) approach to iron out solution bugs before implementing with the rest of the organization

- Fleshing training plans out in more detail

- Additional staff support that may be needed when executing transition activities

- Maintaining communications among staff during the transition process

- Dependencies between Initial Win projects and/or process solutions that are being rolled out.

- Development of a transition schedule

As you look at each solution you will be implementing, consider any one of these strategies based on your specific needs and risks. It is also okay to mix and match many of the strategies previously described.

ITSM Organization Structures

The world would be a much better place if there was a single organization structure that would work for everybody. Unfortunately, this is not the case. The right answer for your organization will vary based on specific company culture, management style, current organization, geographic locations and a number of other factors.

ITSM Organizational Design Considerations

There are some organizational principles that you may wish to consider when defining your specific ITSM Organization approach.

A basic approach is to make sure that there are at least 3 defined reporting structures in place. This might look something like the following:

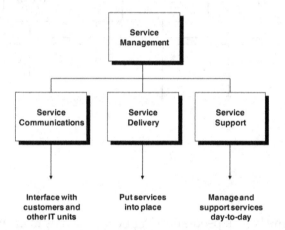

With the above, there are three distinct groupings of functions in the organization. The Service Support function manages services day-to-day. The Service Delivery function plans and builds services to meet business need. The Service Communications function provides an outer interface to the business. It is in this last function that things such as the Service Desk or liaisons to the business may reside.

A simple view of the above would show that the Service Communications function would gather business requirements and provide help and guidance to customers and end-users. The Service Delivery function would then translate business requirements into needed IT solutions. The Service Support function would manage and operate those services day-to-day.

This is a simple view, but highly effective. When reviewing various IT support organizations, look for signs that a similar type of structure is in place. Although some organizations may have complex structures, try to make sure that at some point the organization comes together like that shown above. Experience has shown that organizations that do not organize in a manner close to the above are usually in a state of disarray.

What do you do if your current organization is so focused on technical silos, that something like the above is very far removed from where you would like to be? If that is your situation, you may want to consider a transitional organizational solution. The following has worked with these kinds of situations:

Step #1

Create a transitional ITSM organization that exists as a virtual organization with indirect reports into your existing structure. It might look something like this:

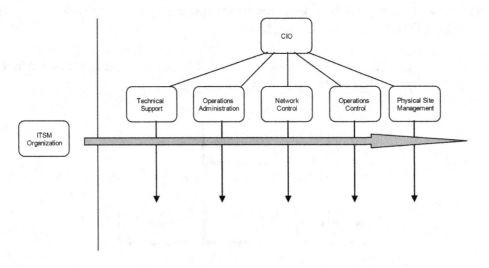

Step #2

After a period of time has elapsed allowing new practices to embed into the culture and operation, you can then "flip" the structure so that it eventually becomes something like this:

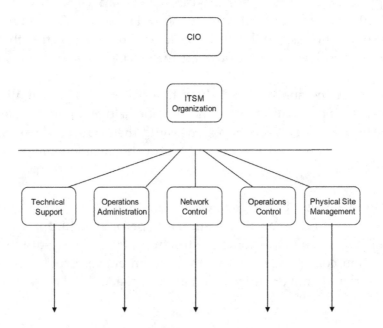

When designing an ITSM Organization Structure, there are typically three main models that most organizations adhere to. They are:

- Centralized

- Distributed

- Workgroup

Key criteria for selecting a model will depend on three factors unique to your organization. The table below summarizes these:

ITSM Organization Criteria

Criteria	Description	Options
Control	Identifies where decisions are made	Single point of control Multiple points of control
Execution	Identifies where ITSM processes will be executed	Single point of execution Multiple points of execution
Communication	Identifies how information will flow throughout the organization	Vertical communication from management to subordinates Peer-to-peer communication between management entities

There is no one correct answer for which model to apply. Each of the models may be chosen based on factors unique to your company's way of doing business. The following describes each of these models.

Centralized Organization Model

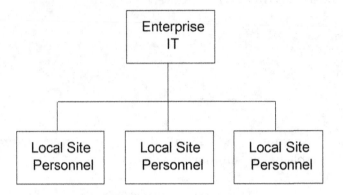

Description:

In this model, all decision making, management and ITSM solution design is centralized in the company. Local sites only execute ITSM activities under the guidance and management of the centralized IT organization.

Characteristics:

Criteria	Description
Control	Single point of control
Execution	Single point of execution
Communication	Vertical communication flow

Situations Best Used in:

- Small Companies
- Companies requiring high degree of control
- Companies that prefer to keep design skills centralized
- Preservation of a dictatorship culture

Considerations:

- Local sites are relieved of management responsibilities
- This model tends to meet the majority of business needs
- Decisions can be made more efficiently and quickly

Distributed Organization Model

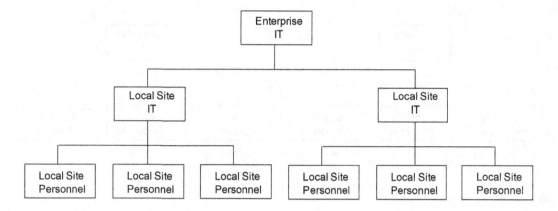

Description:

In this model, IT decision making and solution design is still centralized at the enterprise level. Management and local site autonomy is now distributed in the company. Local sites both manage and execute ITSM activities under the policies and guidance of the centralized Enterprise IT organization.

Characteristics:

Criteria	Description
Control	Single point of control
Execution	Multiple points of execution
Communication	Vertical communication flow

Situations Best Used in:

- Global companies that require global standards and control
- Large companies requiring high degree of control with many local sites
- Companies that prefer to keep design skills centralized

Considerations:

- Allows for distributed decision making and management without losing control
- Places management of ITSM processes closer to those that execute them
- Design and policy decisions can be made more efficiently and quickly

Workgroup Organization Model

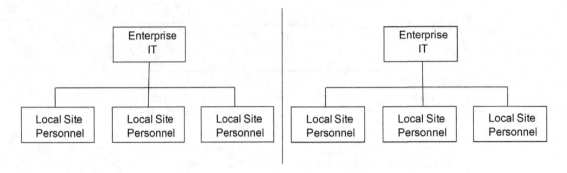

Description:

In this model, all IT decision making, solution design and management is done independently by each processing site. Local sites only execute ITSM activities under the policies and guidance of the local Enterprise IT organization.

Characteristics:

Criteria	Description
Control	Multiple points of control
Execution	Multiple points of execution
Communication	Vertical communication flow

Situations Best Used in:

- Merger situations where there is no interest in combining IT functions
- Global companies where IT organizations and standards are to remain unique to each geography
- Companies where business units may need to operate independently
- Companies that will only obtain ITSM benefits within local geographies

Considerations:

- Works in situations where a single management model cannot be applied
- Will result in redundancies of solutions and designs
- Design, management and policy decisions can be made more efficiently at local sites

ITSM Governance Considerations

The mission of ITSM Governance is to manage the design, deployment, maintenance, and evolution of IT Service Management solutions and ensure an ongoing service improvement lifecycle is maintained. To accomplish this, ITSM Governance will work in partnership with key stakeholders and the ITSM Steering Committee to continually align IT Service Management process solutions with company business strategies.

Employment of an ITSM Governance function should not be overlooked. It plays a vital role in:

- Meeting the goals of creating global process standards with approved local variations where needed to meet business objectives

- Avoiding confusion and discord and inefficiencies created by multiple process approaches while maintaining a consistent customer service experience

- Providing an independent channel and single point of contact for resolving ITSM solution design conflicts

- Managing key process design changes over time as processes mold and adapt to meet the needs of the business

- Coordinating research into new technologies and best practices for IT Service Management that may provide value to the business organization

A number of guiding principles can be used when putting your ITSM Governance solution together. These include principles such as the following:

- Process ownership will be centrally managed through Process Owners

- Process models and documentation will be managed as Configuration Items under the control of Change Management

- Processes will be built and designed globally with allowance for local variation when approved

- The ITSM Steering Committee will be the final decision making authority for IT process changes and will own the governance process

- A process exception and appeals process will be in place to resolve ITSM Program issues and concerns

- Implementation of ITSM solution changes will utilize the Initial Wins Project Lifecycle

- A process assurance group will be in place to assess the purpose and fit of processes and process changes from a regional and local perspective

- Key Performance Indicators (KPIs) will be established, monitored and reported on to assess the overall health of the governance process

A successful ITSM Governance function can be identified by having:

- Clear responsibility for the management of IT processes by Process Owners

- Active sponsorship and championship of IT Process Governance activities by the ITSM Steering Committee

- Proactive stakeholder and business involvement throughout the ITSM Governance Process

- Clear definition of IT Process Governance conformance and compliance standards

- Clear decision making authority and boundaries for Process Owners and the ITSM Steering Committee

- Ongoing education and communication of IT Service Management and fostering an IT service culture throughout the company

Possible metrics you may wish to include with this function are:

- Number of process changes requested
- Number of process exceptions requested

- Number of process changes rejected

- Number of process changes approved

- Number of process changes escalated to ITSM Steering Committee

- Number of process exceptions rejected

- Number of process exceptions approved

- Number of process exceptions escalated to ITSM Steering Committee

- Number of process exceptions escalated to line of business

- Number of major process changes

- Number of minor process changes

- Number of incidents caused by process issues

- Number of incidents related to non-standard processes

There are four main sub-processes that make up the ITSM Governance function. These can be described as follows:

ITSM Governance

This sub-process provides a structured approach for reviewing and approving decisions for ITSM changes to be made in accordance with company standard IT management solutions.

ITSM Exceptions and Appeals

This sub-process provides a means of escalating ITSM decisions for the use of non-conforming solutions to meet unique and/or local business requirements.

ITSM Vitality

This sub-process provides a way to incorporate new ITSM solutions and changes into the company ITSM standard as a result of changing business needs. This sub-process also researches new ITSM solutions and best practices in the marketplace.

ITSM Communications

This sub-process provides a means for syndicating ITSM solutions as they evolve across the business enterprise. Personnel involved in Organizational Change activities may also have a home here on an ongoing basis.

An example of how these fit into an overall ITSM Governance Model can be shown as follows:

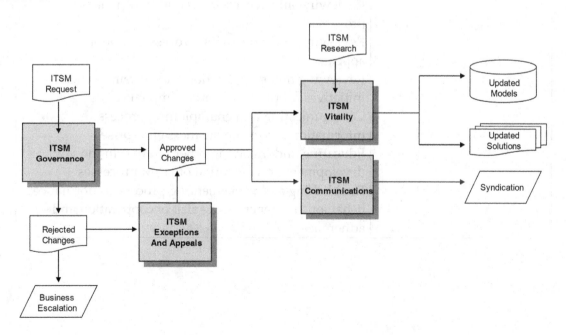

The following presents suggested roles to operate the ITSM Governance Process:

ITSM Governance Roles

Governance Role	Key Activities
ITSM Steering Committee	Sponsoring and championing ITSM solutions across the company enterprise Reviewing and approving major ITSM changes Reviewing and providing feedback on the effectiveness of ITSM solutions and governance
Service Manager	Providing leadership and guidance to Process Owners Liaison to the ITSM Steering Committee Ensuring business requirements are being addressed by ITSM activities Coordinating appropriate resources for key ITSM initiatives Scheduling and coordinating solution meetings and key activities Acting as first level escalation point for resolving ITSM exceptions and appeals
Process Owner	Developing and updating processes and process models Managing, executing and maintaining processes Ensuring business requirements are supported by the ITSM processes Monitoring technical changes being implemented and their impact to processes Reviewing and processing requested changes to processes Reviewing and processing process exception and appeals Reviewing major application and system services initiatives for process fit and compliance Communicating and championing process information across the business enterprise Educating and involving other IT areas in the development and evolution of ITSM processes Establishing and implementing process conformance behavior that generates a spirit of cooperation and adherence

Governance Role	Key Activities
ITSM Stakeholder	Understanding the ITSM Governance strategy Locally championing and supporting ITSM solutions Reviewing ITSM changes as necessary and providing feedback to Process Owners Leveraging ITSM solutions for reuse and conformance to the greatest extent possible when building or implementing IT solutions Submitting ITSM change requests and working in a cooperative manner to resolve solution conflicts Escalating ITSM concerns and issues as necessary
ITSM Governance Analyst	Understanding the ITSM Governance strategy Locally championing and supporting ITSM solutions Reviewing ITSM changes as necessary and providing opinions and recommendations to ITSM Steering Committee Providing feedback on ITSM solution fit to Process Owners and Process Stakeholders

The following pages identify example workflows and role responsibilities for the ITSM Governance function.

ITSM Governance Workflow Model

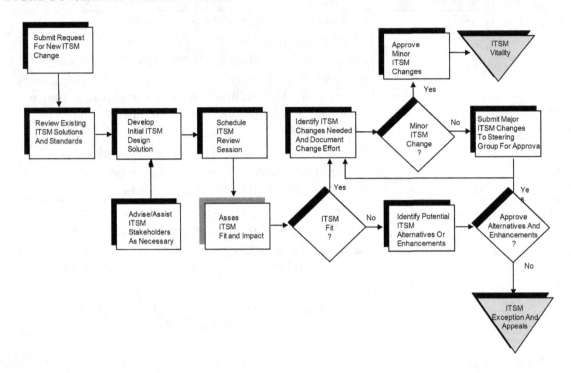

Role Responsibilities:

Governance Activity	ITSM Stakeholder	Service Manager	Governance Analyst	Process Owner	Steering Group
Submit request for new ITSM change	x			x	
Review existing ITSM solutions and standards				x	
Develop initial ITSM design solution				x	
Schedule ITSM review session(s)				x	
Assess ITSM solution fit and impact		x	x	x	
Identify solution alternatives/enhancements		x	x	x	
Identify ITSM changes needed and document change effort				x	
Approve minor ITSM changes				x	
Submit major ITSM changes to Steering Group for approval				x	
Advise/Assist process stakeholders as necessary				x	
Submit request for new process change				x	
Review existing processes and standards				x	

ITSM Exceptions and Appeals Workflow Model

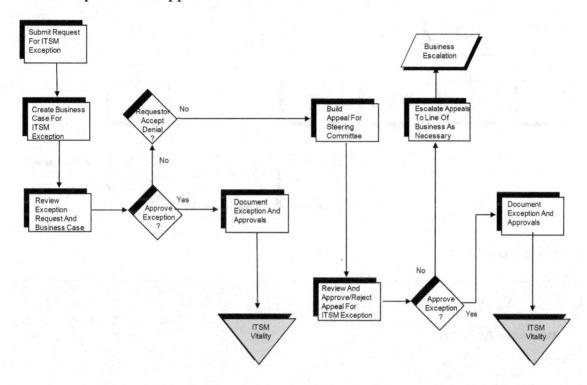

Role Responsibilities:

Governance Activity	ITSM Stakeholder	Service Manager	Governance Analyst	Process Owner	Steering Group
Submit request for ITSM exception	x				
Create business case for ITSM exception	x				
Review exception request and business case		x	x	x	x
Build appeal for Steering Committee	x	x		x	
Review and approve/reject appeal for ITSM exception		x	x	x	x
Escalate appeals to line of business if necessary	x				
Document exception and approvals				x	

ITSM Vitality Workflow Model

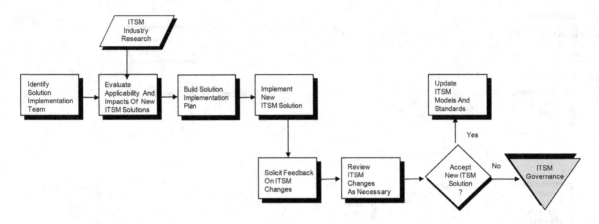

Role Responsibilities:

Governance Activity	ITSM Stakeholder	Service Manager	Governance Analyst	Process Owner	Steering Group
Identify solution implementation team				x	
Evaluate applicability and impacts of new ITSM solutions		x	x	x	
Build solution implementation plan		x	x	x	
Implement new ITSM solution				x	
Solicit feedback on ITSM changes				x	
Review ITSM changes as necessary		x	x	x	
Update ITSM models and standards				x	

ITSM Communications Workflow Model

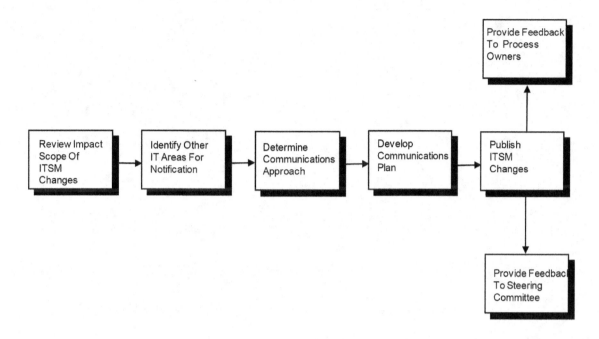

Role Responsibilities:

Governance Activity	ITSM Stakeholder	Service Manager	Governance Analyst	Process Owner	Steering Group
Review impact scope of ITSM changes				x	
Identify other IT areas for notifications		x	x	x	
Determine communications approach				x	
Develop communications plan				x	
Publish process changes				x	
Provide feedback to Process Owners		x			
Provide feedback to Steering Committee				x	

ITSM Tooling Considerations

"Do not lie in a ditch, and say God help me; use the lawful tools He hath lent thee."
—English Proverb

The following pages present a number of high level considerations for tools that can be used to support ITSM activities. The descriptions presented here are generic, in that vendor specific products are not specifically referred to. Each tool is described in terms of generic requirements. This is useful as a start for examining potential vendor tools.

Vendor tools sold in the marketplace will typically fall into many of the categories shown. The goal here is to present an overall holistic view of the kinds of tools needed and generic requirements. Vendor tool proposals can then be compared against this. Some tools may map one-for-one against the types shown. Others may include more than one tool type or, in some cases, only provide some of the functionality shown for a given type.

The pages following describe a full set of general requirements needed for IT Service Management. This includes a reference chart that indicates which ITSM process is supported by which kinds of tools.

A logical data model is also presented with high level data components for each ITSM database.

Summary of ITSM Tools by Process

ITSM Tool Type	Service Desk	Incident	Problem	Change	Release	Configuration	Service Level	Availability	Capacity	Svc. Continuity	Financial	Security	Other
Asset Management Database	X					X				X	X		X
Auto Discovery Tool					X	X						X	X
Backup and Restore Tool										X			X
Automatic Call Distribution (ACD)	X												
Availability Management Database								X					
Call Center Staffing Calculator	X												
Call Center Telephony Infrastructure	X												
Capacity and Performance Monitor		X						X	X				X
Capacity and Performance Simulation Tool					X				X				X
Capacity Management Database									X				
Capacity Modeling Tool									X				
Change Management Database	X	X	X	X	X	X	X	X	X	X	X	X	X
Change Management Workflow Manager	X	X	X	X	X	X	X	X	X	X	X	X	X
Chargeback Tool											X		
Configuration Management Database	X	X	X	X	X	X	X	X	X	X	X	X	X
Definitive Software Library (DSL) Manager				X	X	X							
Distribution List Manager	X	X	X	X	X	X	X	X	X	X	X	X	X
Documentation Support Tool		X	X	X	X	X	X	X	X	X	X	X	X
Event Correlation Tool	X	X	X					X	X			X	X

ITSM Tool Type	Service Desk	Incident	Problem	Change	Release	Configuration	Service Level	Availability	Capacity	Svc. Continuity	Financial	Security	Other
Financial Management Database							×	×	×	×	×		
Forward Change Calendar	×	×	×	×	×	×	×	×	×	×	×	×	×
Identity Manager					×	×						×	×
Incident Management Database	×	×	×		×		×	×	×			×	×
Incident Management Knowledge Base	×	×	×		×			×	×			×	×
Incident Management Workflow Manager	×	×										×	×
Interactive Voice Response (IVR)	×				×								
Intrusion Detection Monitor		×										×	×
IT Service Continuity Management Database										×			
Performance Analysis Tool									×				×
Problem Management Database	×	×	×		×		×	×	×			×	×
Reader Board	×												×
Release Build Manager				×	×								×
Release Management Database				×	×			×	×	×			×
Release Test Manager					×								×
Remote Support Tool	×	×	×		×		×	×	×	×		×	×
Report Generator	×	×	×	×	×	×	×	×	×	×	×	×	×
Scheduler					×								×
Security Access Manager												×	
Security Monitor		×										×	×
Service Continuity Planning Tool										×			

ITSM Tool Type	Service Desk	Incident	Problem	Change	Release	Configuration	Service Level	Availability	Capacity	Svc. Continuity	Financial	Security	Other
Service Level Management Database	X	X	X	X	X	X	X	X	X	X	X	X	X
Service Management Dashboard	X	X					X	X	X	X			X
Service Management Workflow Manager							X	X	X	X	X	X	
Software Configuration Manager					X	X							
Software Distribution Manager					X								
System Directory	X				X			X	X			X	X
System Event Monitor	X	X	X									X	X
Test Data Generator					X								X

Generic Tooling Requirements

This section describes each tool type listed in the previous summary table along with key functional requirements that should be included with that tool type.

Asset Management Database

The key functions of this tool type should include:

- Storage of information about CI assets
- Linkage to Configuration Management databases
- Support for Asset status and lifecycle management such as procurement, stored, configured, deployed, active and retired stages to support release impact analysis, planning, rollout and deployment activities
- Ability to track a wide variety of enterprise IT asset types (hardware, software and services)
- Ability to record a wide variety of contracts and licensing agreements
- Ability to track the physical location of contracts and agreements, and identify the individuals responsible for them
- Ability to interface data with and from automated discovery applications
- Ability to group an individual customer's/user's assets and services to provide cost information
- Ability to generate reports based on geography, user, CI type and impact
- Ability to perform ad hoc/general queries
- Ability to integrate with other ITSM databases and applications
- Ability to manage leases, depreciation schedules, warranties, and service provider contracts as critical elements
- Ability to track both fixed and variable costs of IT assets
- Ability to tie IT assets to related items (e.g., individuals, contracts, leases, warranties, service-level agreements, other assets, incidents, problems)
- Integration with ITSM Configuration Management Database
- Ability to support a web-based front end that can significantly reduce the amount of administrative overhead
- Ability to manage moderate-to-large quantities of complex IT asset data using a standard, non-proprietary database engine
- Ability to support both flexible data import/export, and simple points of integration for associated tools

Auto Discovery Tool

The key functions of this tool type should include:

- Automated checks for sufficient available disk space, hardware and software prerequisites in the live environment to ensure compatibility for new or changed releases

- Hardware and software audit reporting

- Flexible ability to filter, add or limit audit items being checked for or to match against a defined set of configuration requirements

- Automatic event notification for workstations or servers that fail audits

- Hardware and software audit reporting

- Flexible ability to filter, add or limit audit items being checked for or to match against a defined set of configuration requirements

Backup and Restore Tool

The key functions of this tool type should include:

- Ability to backup / restore locally and over the network
- Ability to backup network operating system specific files
- Ability to backup system files
- Ability to communicate to user status of file, location of data for recovery
- Ability to do both full and incremental backups
- Ability to generate archive cycle list
- Ability to generate recovery scripts
- Ability to handle data compression
- Ability to identify logical groupings of files
- Ability to manage backup media
- Ability to operate across different platforms
- Ability to perform open file backups
- Ability to point in time restores
- Ability to provide problem notification
- Ability to restore alternate path
- Ability to restore archived data
- Ability to restore back to specific time
- Ability to restore by logical groupings of files
- Ability to restore cross platform
- Ability to restore from incremental and full backups
- Ability to restore individual files from full backup
- Ability to restore system files
- Ability to support and modify retention rules
- Ability to support multiple media
- Ability to verify success / failure of backup/restore activities
- Access to an archive and historical database
- Access to central repository for archive / historical database
- Catalog of media location / status
- Provide command line and GUI interfaces
- Support high volumes in short periods of time
- Search engine to find and call up media
- Support for hot backups

Automatic Call Distribution (ACD) Tool

The key functions of this tool type should include:

- Process calls on a first-come first-serve basis

- Answer all calls immediately

- Place callers in call-queue and direct to next available Call Agent

- Support multiple call queues

- Log call group activity

- Provide capability to monitor call group activity, analyze call queues and Call Agent hold time

- Provide different call processing paths based on defined business rules

- Provide capability for calling route priority handling

- Allow for establishment of thresholds to minimize delays and route calls to alternative queues

- Allow for routing of calls based on incoming phone number dialed (DNIS)

- Provide incoming line capacity adequate to support the desired infrastructure

- Interface with corporate telephone and communications systems

- Manage incoming call traffic by employing a queuing or waiting list assignment to each call

- Provides management information reports on agent performance

- Provides real time monitoring of workload

- Provide capability to route to a recorded message and then connect to the first available agent if all agents are busy

- Provide logging capability for call management statistics

- Provide interfaces to Incident and Financial Management databases

Availability Management Database

The key functions of this tool type should include:

- Allow for collection of data from diverse sources including systems, networks, applications, web servers and phone systems

- Provide a Data Warehouse capability for storage of availability related information

- Provide storage of historical availability plans, risk assessments and reports

- Data can be archived to tape and restored as needed

- Provide interfaces to ITSM repositories such as Incident, Problem, Service Level, Capacity and Financial databases

- Provide interfaces to System and Security Monitors and Event Correlation Tools

- Components may be customized to meet specific requirements related to the extraction, integration, storage and management of system availability from multiple sources

- Data access extends across multiple hardware platforms and operating systems, networks, web servers, telephone systems, and includes statistics generated by custom applications

- Interface with statistical calculation and summarization tools (i.e. SAS, SPSS, etc.)

Call Center Staffing Calculator

The key functions of this tool type should include:

- Capability to estimate number of staff required to provide Service Desk call agent services based on forecasted call volumes, estimated call durations and productive employee time per day

- Provide indication of staffing shortages

- Estimate call abandon times based on input call volumes and number of agents

- Provide what-if scenarios for modeling staffing plans and key events that may impact planned call volumes

- Provide capability for storage of multiple modeling scenarios

Call Center Telephony Infrastructure

The key functions of this tool type should include:

- Provide capability for toll-free access

- Provide telephone headsets or other hands-free devices

- Interface with call management systems

- Capability to place callers on hold with ability to contact others without losing the caller

- Provide adequate number of telephone stations to handle required staffing levels plus emergency overloads

Capacity and Performance Monitor

The key functions of this tool type should include:

- Gather key metrics for the hardware components (mainframe, network, server, and workstation):—utilization, throughput, workload, buffers, response

- Gather key metrics for the software components (system, middleware, application): utilization; throughput; buffers; queues; response

- Consolidate metrics by time, group of components

- Ability to interface to standard corporate desktop

- Use standard event protocol to communicate event status

- Ability to interface with System Event Monitor

- Capability to run scripts including unattended

- Capability to send alarms/alerts

- Ability to measure/determine idle cycles

- Ability to capture data by machine

- Ability to capture data by process

- Ability to correlate cause and effect analysis

- Ability to work in exception more

- Access to application configuration info

- Access to historical component database

- Centralized, real-time, and historical trending monitoring capability

- Collect and consolidate data into central repository—organize by CI

- Consolidate capacity data from individual components

- Supports start and stop criteria

- Trace hooks (API) for applications

- Web interface

- Ability to record task ownership and restrict access by owner

- Common front end for parameter setting

Capacity and Performance Simulation Tool

The key functions of this tool type should include:

- Capability to model and simulate application solutions before implementing them in a production environment reducing costly and time consuming test cycles

- Capability to evaluate application flow and logic, discover and remove bottlenecks, and predict needed performance and capacity

- Capability to track state and logic of applications as they execute in test environment to identify potential performance bottlenecks

- Logging function to track recorded performance and timing information

- Ability to simulate transaction volumes

- Capability to model complex changes in application path based on business logic or other rules

Capacity Management Database

The key functions of this tool type should include:

- Allow for collection of data from diverse sources including systems, networks, applications, web servers and phone systems

- Provide a Performance Data Warehouse capability

- Provide storage of historical capacity models and reports

- Provide capability to automatically reduce data over time to conserve disk resources, while enough data is retained for baseline, trend and exception analysis

- Data can be archived to tape and restored as needed

- Provide interfaces to ITSM repositories such as Incident, Problem and Financial databases

- Components may be customized to meet specific requirements related to the extraction, integration, storage and management of system performance data

- Data access extends across multiple hardware platforms and operating systems, networks, web servers, telephone systems, and includes statistics generated by custom applications

- Interface with statistical calculation and summarization tools (i.e. SAS, SPSS, etc.)

Capacity Modeling Tool

The key functions of this tool type should include:

- Predicts process, transaction and job response times based on system loads and volumes

- Predicts hardware utilization levels based on system loads and volumes

- Allows for correlation between specified business workloads and volumes with system workloads and volumes

- Allows for what-if scenario modeling including changes in hardware and software platforms as well as workload volumes

- Predicts application performance under varying loads

- Allows for prediction for unexpected spikes in demand

- Identifies performance bottlenecks within the infrastructure being modeled

- Provides workload and queue result statistics both summarized and in detail

- Interfaces to reporting tools for creation of capacity forecasts

- Provides flexibility in modeling such as standard M/M/1 queuing or stochastic modeling approaches

- Allows for storage of multiple versions of models

- Provides capabilities to create baseline models and compare results to existing performance

- Provides capabilities to compare what-if scenarios to baseline models

- Provides flexible resource modeling components based on actual vendor platforms

- Allows for storage of multiple versions of models

- Provides capabilities to create baseline models and compare results to existing performance

- Provides capabilities to compare what-if scenarios to baseline models

- Provides flexible resource modeling components based on actual vendor platforms

Change Management Database

The key functions of this tool type should include:

- Repository to store, retrieve and report on RFCs in an easily accessible format

- Capability to link RFCs to projects

- Links to the Configuration Management database or Configuration Management data

- Means to identify easily the other CIs that will be impacted whenever a Change to any specific CI is proposed

- Ability to allow Change Management staff, Change builders, testers, etc. to add text to Change records

- Ability to store back-out procedures should a Change cause problems

- Ability to store historical reports on Change Management activity

Change Management Workflow Manager

The key functions of this tool type should include:

- Automatic production of requests for impact and resources assessment to the owners of the impacted CIs

- Ability for all authorized personnel to submit RFCs from their own terminal or location

- Ability to progress requests through the appropriate stages of authorization and implementation and to maintain clear records of this progress

- Automatic warnings of any RFCs that exceed pre-specified time periods during any stage

- Automatic prompting to carry out reviews of implemented Changes

- Ability to build Changes

- Automatic production of Forward Schedule of Changes (FSCs)

- Process/workflow feature

Chargeback Tool

The key functions of this tool type should include:

- Integrate with Capacity Management Database for usage information

- High flexibility provided for development of charging algorithms based on complex business logic

- Can allocate IT resources to specific cost centers

- Able to apply unique rates to a variety of IT services

- Support various billing modes such as zero-balancing, pro-ration and others

- Provide support for corporate audit functions

- Support various invoicing calendars

- Provide multi-currency support

- Provide capability to mix charge-back methodologies, and to easily apply changes in response to new requirements

- Provide capability for business units to track, analyze and manage their own usage and costs

- Effectively support and manage billing and invoicing cycles

Configuration Management Database

The key functions of this tool type should include:

- Security controls to limit access on a need-to-know basis

- Support for CIs of varying complexity e.g. entire systems, Releases, single hardware items, software modules, or hierarchic and networked relationships between CIs; by holding information on the relationships between CIs, Configuration Management tools facilitate the impact assessment of RFCs

- Easy addition of new CIs and deletion of old CIs

- Automatic validation of input data (e.g. are all CI names unique)

- Automatic establishment of all relationships that can be automatically established, when new CIs are added

- Support for CIs with different model numbers, version numbers, and copy numbers

- Automatic identification of other affected CIs when any CI is the subject of an Incident report/record, Problem record, Known Error Record or RFC

- Integration of Problem Management data within the CMDB, or at least an interface from the Configuration Management system to any separate Problem Management databases that may exist

- Automatic updating and recording of the version number of a CI if the version number of any component CI is changed

- Maintenance of a history of all CIs (both a historical record of the current version—such as installation date, records of Changes, previous locations, etc.—and of previous versions)

- Support for the management and use of configuration baselines (corresponding to definitive copies, versions etc.), including support for reversion to trusted versions

- Ease of interrogation of the CMDB and good reporting facilities, including trend analysis (e.g. the ability to identify the number of RFCs affecting particular CIs)

- Ease of reporting of the CI inventory so as to facilitate configuration audits

- Flexible reporting tools to facilitate impact analyses

- Ability to show graphically the configuration or network maps of interconnected CIs, and to input information about new CIs via such maps

- Ability to show the hierarchy of relationships between parent CIs and child CIs

Definitive Media Library (DML) Manager

The key functions of this tool type should include:

- Capability to store and manage authorized versions of media in a secured environment

- Provide media license management capabilities to track usage of media and compliance with media contracting and purchase agreements

- Provide support for external media license audits

- Provide check-in/check-out facilities for media with access notification and status

- Provide secure access in line with corporate security policies

- Provide interfaces with Configuration, Change and Release Management databases

- Provide linkages with IT Service Continuity Management to ensure authorized media will not get lost in the event of a major business disruption

- Provide support for media version control and naming/numbering standards

Distribution List Manager

The key functions of this tool type should include:

- Storage and maintenance of notification distribution lists to support ITSM activities such as Incident, Problem and Service Level Management

- Provide interfaces to company e-mail applications and services

- Flexible support for both individual as well as group notifications

- Linkages to Systems Directory

- Support for corporate security policies

Documentation Support Tool

The key functions of this tool type should include:

- Provide on-line and/or web access to documentation

- Provide access to Company policies, guidelines, and procedures

- Provides access to Company design requirements

- Ability to provide multi user access

- Ability to provide flexible authorization and authentication

- Access to organizational information regarding availability or general standards

- Provide search engine capability to find documentation

- Support for documenting processes and procedures

- Hypertext capability to reference associated documents

- Ability to interface to standard Corporate desktop

- Support version control and compliance with Corporate Documentation standards

Event Correlation Tool

The key functions of this tool type should include:

- Ability to correlate events from different sources based on rules

- Ability to provide flexible and dynamic correlation rules

- Ability to analyze all received events

- Ability to recognize and filter repetitive events

- Ability to analyze events based on arrival time stamp

- Ability to modify event data

- Ability to accept events from different sources and protocols

- Ability to filter and threshold event data

- Ability to generate events

- Interface to Configuration Management Database

- Ability to utilize other information sources such as configuration data, i.e. ability to compare correlation tool's configuration information with actual discovered configuration information

- Ability to Initiate actions based on rules

- Ability to interface with Incident Management Database

- Ability to automatically log an Incident

- Ability to interface with other tools with open APIs

- Ability to review the Change Management Forward Schedule of Changes to determine if event from a scheduled outage/change

- Ability to send notifications in multiple formats

- Provide multi user access

- Provide flexible authorization and authentication

- Provide local and remote access

- Not restricted to a single operating system

- Ability to access event information locally and remotely

Financial Management Database

The key functions of this tool type should include:

- Provide capabilities for graphical reporting and multi-dimensional analysis of costs and resources

- Provide support for internal and external audit activities

- Provide flexible data analysis functions and features

- Provide storage for historical IT financial reports

- Provide storage of historical customer bills

- Provide storage and support for IT budgets and cost models

- Provide interfaces to corporate accounting infrastructure and general ledger

Forward Change Calendar

The key functions of this tool type should include:

- Ability to link RFCs to forward change events and schedules

- Linkage with Change Management Database and Distribution List Management

- Calendaring capabilities that can be viewed by those requesting changes as well as those administering and implementing changes

- Ability to easily reschedule changes and identify scheduling conflicts

Identity Manager

The key functions of this tool type should include:

- Provide self-service and password reset/sync interfaces

- Provide access via web services

- Interface with corporate security policies

- Provide centralized control and local autonomy functions

- Provide API interface to applications

- Workflow engine for automated submission and approval of user requests

- Provisioning engine to automate the implementation of administrative requests

- Automatic synchronization of user data from different repositories like human resources databases and enterprise directories

- A sophisticated role-based administration model for delegation of administrative privileges

- Role and rule-based delegated administration

- Intelligent approval routing

- Auditing and reporting mechanisms

Incident Management Database

The key functions of this tool type should include:

- Repository to store, retrieve and report on Incidents in an easily accessible format

- Capability to link to Problem Records and RFCs related to Incidents

- Links to the Configuration Management database or Configuration Management data

- Capability for storing historical incident data and other Incident related information

- Capability to store Incident Management report data and historical reports

- Ability to allow those involved with Incident Management activities to add text to Incident records

- Ability to store and maintain alerting distribution lists

- Flexible support for desired Incident classification and logging schemas

- Flexible search capabilities

Incident Knowledge Base

The key functions of this tool type should include:

- Storage and retrieval of resolution and recovery actions to avoid repeating those actions again for similar Incidents

- Highly flexible search capabilities

- Maintenance utilities to maintain knowledge over time

- Linkages to CMDB, Incident Database, Distribution lists and other repositories that can provide further actions or detail as needed to research Incidents

Incident Management Workflow Manager

The key functions of this tool type should include:

- Highly flexible routing of Incidents across control staff taking resolution and recovery actions

- Flexible support for Incident workflow activities throughout the Incident lifecycle

- Support for control staff that may be located in multiple sites or co-located in an operations bridge

- Automatic escalation facilities so as to facilitate the timely handling of Incidents and service requests

- Support and coordination for escalation and approval checkpoints

- Ability to time stamp workflow actions as they occur

- Ability to automatically issue notification alerts based on Incident type and user groups

- Linkages to Distribution List Manager

- Flexibility to tie distribution lists and notification types to Incident categories

- Automatic Incident logging and alerting in the event of fault detection on mainframes, networks, servers (possibly through an interface to system management tools)

- Automatic modifications to the Incident record being registered in order to keep control

- Linkages with Incident Management database

Interactive Voice Response (IVR) Tool

The key functions of this tool type should include:

- Play recorded messages including information extracted from databases and the internet

- Route calls to Call Agents based on programmed business logic

- Provide capability to transfer callers to outside extensions

- Provide capability to collect and store caller provided information before call is transferred to a Call Agent

- Provide capability to fulfill and complete caller requests without a transfer

- Provide capability to verify caller identity before transferring the call

- Intelligently route calls using complex business logic

- Interface with corporate telephone and communications systems

Intrusion Detection Monitor

The key functions of this tool type should include:

- Ability to drive detection by security policy

- Support protection against intrusions that could result in legal liability

- Support protection against confidentiality breaches

- Support protection against email imbedded viruses

- Provide real-time protection against spy ware

- Provide automatic removal of breaches and infections for easy disposal of security risks

- Provide support for cleanup of registry entries, files, and browser settings after spy ware intrusion detected

- Support protection against spam attacks

- Support protection against degradation and loss of network services through misuse and hostile attacks

- Support protection against infection and loss of data from Web-based viruses

- Support protection against corruption of data from malicious code embedded in programming tools and scripts (i.e. Java, ActiveX, etc.)

- Prevent access to inappropriate sites by internal staff

- Provide capability to set spy ware and ad ware policies on an individual application-by-application basis

- Provide timely and periodic detection updates to stay current with latest intrusion threats

- Provide intrusion scanning facility that can be scheduled or manual

- Minimize work disruptions and slowdowns caused by scanning activities

IT Service Continuity Database

The key functions of this tool type should include:

- Provide storage and support for IT Service Continuity plans as well as plan versions

- Provide support for storage and management of IT Service Continuity test plans, test data and test results

- Interface with corporate Business Continuity Management data repositories

Performance Analysis Tool

The key functions of this tool type should include:

- Analyzes performance and operational data from different CI types

- Analyzes service chains of CIs to obtain end-to-end performance

- Interfaces to Service Level Management tools to track performance against established service targets

- Provides real-time and historical analysis of performance data and trends

- Allows for drill-down capabilities to users, CIs and processes active at the time of a performance problem

- Provides support for exploration of cause-and-effect relationships

- Uncovers cycles and patterns in system behavior

- Provides alerting mechanisms to notify staff of potential problems before users are affected

- Provides real-time views with up-to-the-minute performance data

- Interfaces to Capacity Management Database to maintain a historical record of system and application performance

- Identifies trends, cycles and patterns to better understand performance

- Provides trend analysis capabilities to prepare for and avoid performance problems and outages

- Provides interfaces to application programs and developers to allow for performance monitoring within application segments and code

- Provides interfaces to database systems to monitor performance of those systems as well as access calls to them

Problem Management Database

The key functions of this tool type should include:

- Repository to store, retrieve and report on Problems and Known Errors as well in an easily accessible format

- Repositories for the collection of appropriate incident/incident lifecycle data and information

- Capability to link to RFCs related to error removal

- Links to the Configuration Management database or Configuration Management data

- Capability for storing historical incident data and other problem related information

- Capability to store problem management report data and historical reports

- Ability to allow those involved with Problem Management activities to add text to Problem records

- Links to Incident Management database to obtain incident data to assist with problem identification and resolution

- Links to Incident Management database to communicate information on Known Errors to Incident staff

- Ability to link Incident records with Problem records

Reader Board

The key functions of this tool type should include:

- Provide display capability that allows messages to be viewed by technology center staff

- Allow for manual entry of messages

- Interface with System Event Monitors and call management infrastructure

- Provide enhanced display to notify technology center staff based on message severity or impact

- Audible alarm capability for critical incidents

Release Build Manager

The key functions of this tool type should include:

- Supports automated build of new Releases of software applications for mass distribution and use

- Ability to drive program compilations and links, in the correct sequence, under program control using the correct versions of the source code as stored in the DSL

- Ability to generate master CD-ROMs or other media version of software that will be replicated for mass distribution

- Ability to support use of the cross-reference information stored in the Software Configuration Management tool to determine which parent CIs need to be rebuilt when lower-level units are changed

Release Management Database

The key functions of this tool type should include:

- Repository to store, retrieve and report on Releases in an easily accessible format

- Capability to link to Problem Records and RFCs related to Releases

- Links to the Configuration Management database or Configuration Management data

- Capability to store Release Management report data and historical reports

- Ability to store and maintain distribution lists for staff related to release activities

- Flexible search capabilities

Release Test Manager

The key functions of this tool type should include:

- Ability to automate the creation of test data to match test criteria

- Ability to simulate processing volumes and loads for stress testing

- Ability to manage the creation of test criteria and log results against criteria to support test reporting and status

- Ability to simulate end-user keystrokes and/or other actions that would simulate use of the live system

Remote Support Tool

The key functions of this tool type should include:

- Remote diagnostic capability to gather information related to Incidents at other sites

- Ability to monitor user sites as necessary to gain views into symptoms

- Ability to assume control over user terminals, workstations and server operations to assist in Incident resolution and recovery actions

- Real time chat facility

- Secure file transfer capability

- Ability to view remote terminal screens and consoles

- Shared web browsing capability

- Remote-to-local printing without the installation of additional drivers

- Ability to remotely edit registry, schedule reboots, manage users, and administer processes and services

- Ability to see remote CPU/file usage and virtual memory settings

- Support for command line installation and scripted mass-deployment

- Support for two direction file transfer and automatic folder synchronization

- Ability to provide support access and control from handheld devices

- Ability to operate within corporate security policies

- Auto-resolution of dynamic IP addresses for seamless connectivity

- Ability to provide remote support without needing to reconfigure firewalls or other secure perimeter infrastructure

- Remote control of operations for servers and desktops, for example, to assist with making Changes to a server as part of a Release rollout

- Remote monitoring of the event logs and other Problem logs on servers

- Remote monitoring of processor, memory and disk utilization

- Remote management of the disk space on servers—for example to monitor usage, to reorganize files for improved performance, and to allocate more disk space to releases

- Ability to restrict changes that individual Users can make to client workstations to make the target destination for new Releases much more reliable

Report Generator

The key functions of this tool type should include:

- Data search capabilities

- Flexible data extraction capabilities

- Report generation capability

- Statistical analysis on stored data

- Automatic generation of management and trend information relating to Incidents

- Ad-Hoc reporting capabilities

- Flexible data formatting and extraction capability

- Ability accept different input formats

- Ability to provide attachments to reports

- Ability to import/export data using open APIs

- Ability to utilize corporate documentation standards

- Web reporting capability to allow distribution of reports on the corporate Intranet

Scheduler

The key functions of this tool type should include:

- Ability to coordinate all shifts and production work across multiple platforms from a single point of control

- Automate complex and repetitive operator tasks

- Provide capability to dynamically modify production workload schedules in response to changes in the environment

- Manage and resolve workload dependencies

- Manage and track units of processing work

- Monitor processing work for failures and execute pre-programmed restart, backup and recovery processes based on business logic

- Notify operational staff with status of operational schedules

- Provide real-time display of schedule progress

- Provide capability for manual intervention during schedule runs if needed

- Provide capability to model schedules based on estimates in run times or business events

- Provide security functions to protect data and applications

Security Access Manager

The key functions of this tool type should include:

- Prevention of unauthorized access by using a single security policy server to enforce security across multiple file types, application providers, devices and protocols

- Load balancing to prevent performance bottlenecks

- Rules based authorization engine

- Multiple directory support

- Customer Self-Registration Template

- Interface to Security Monitor

- Auditing and reporting capabilities

Security Monitor

The key functions of this tool type should include:

- Raise and escalate security related events to IT staff based on security policy and business logic

- Support for a wide variety of system and networking platforms, devices and file systems

- Detection and logging of file additions, deletions and changes to content, permission and attributes

- Detection of changes to file directories and directory permissions and attributes

- Provide rule base capability for security event detection and escalation

- Automatic archiving capability to store changes in the infrastructure to minimize record keeping and satisfy audit objectives

- Import functionality for external asset and inventory lists to minimize manual entry

- Identification of device configurations highlighting missing configuration parameters

- Enforce separation of duty to ensure that only authorized personnel can designate baselines of the infrastructure

- Interface to Incident Management Database

- Automate scans of the infrastructure to reduce the cost and time associated with manual security checks

Service Continuity Planning Tool

The key functions of this tool type should include:

- Automatic generation of information and population of the IT Service Continuity Plan

- Provide contingency audit and review questionnaires

- Provide framework and checklist for the creation of an IT Service Continuity Plan

- Provide a Business Impact Analysis questionnaire/guide

- Provide Dependency Analysis questionnaire and guide

- Flexibility to make plan updates and changes

- Allowance for multi-user capabilities, allowing cross-departmental collaboration

- Compatibility with corporate platforms and systems

- Recognition of Vital Business Functions as key means for driving the plan

- Interface to ITSCM Database

Service Level Management Database

The key functions of this tool type should include:

- Store SLAs, OLAs and Underpinning contracts

- Capability to support a variety of SLA structures such as master SLA agreements with extensions for unique business unit needs

- Capability to store service data such as SLA/OLA results

- Capability to maintain service historical data and information

- Ability to tie SLA/OLA agreements with business units and departments

- Ability to store service communication distribution lists for service reporting

- Provides linkages between service catalog and SLA/OLAs

- Provide linkage to procurement databases with underpinning contracts

Service Management Dashboard

The key functions of this tool type should include:

- Report or on-line display of services and their delivery status

- Indication as to whether service levels are being met

- Highlights service issues and failures

- Provides different management, service and operational views

- Provides drill-down to sub-services and other information sources

- Linkages to Service Level Management Database

- Interfaces to System Event Monitors and Event Correlation Tools

Service Management Workflow Manager

The key functions of this tool type should include:

- Ability to generate service workflows by SLA/OLA and Service Catalog categories

- Support for workflows to build, agree, approve and maintain SLA/OLAs

- Linkages to Service Level Management Database

- Support for multiple generations of service agreements

- Flexible support for service agreement structures such as a base service agreement with unique business unit addendums

- Logging of negotiation and contact status

- Integrated to-do lists with calendaring

- Flexible maintenance support for SLAs, OLAs and UCs

- Integration with ITSM Service Catalog

Software Configuration Manager

The key functions of this tool type should include:

- Manages the different versions of software source code during its development

- Manages relationships between software components to identify software CI changes for impact upon other parts of the release

- Flexible and configurable support for release packaging requirements such as delta, full and packaged release units

- Linkage with Change Management and Problem Management databases to link software CIs and release units back to a change or problem

- Supports automatic assembly of release components for compile, build, link-edit to executables

Software Distribution Manager

The key functions of this tool type should include:

- Assured delivery of software files

- Integrity checking of data sent and the ability to restart broken transmissions from the point of failure

- Variety of delivery options to optimize the usage of network capacities

- Courier mechanism via CD-ROM, for sending whole packages to remote installations not part of a network

- Fan-out capabilities to employs intermediate servers at remote locations to help with distribution activities

- Ability to store a new version of an application in a dormant state to support rapid activation or deactivation when triggered

System Directory

The key functions of this tool type should include:

- Provide a single point of management for all infrastructure user accounts, devices and applications

- Provide an authorized place to store information about network-based entities, such as applications, files, printers, and people

- Manage identities and broker relationships between distributed resources

- Interface with management and security mechanisms

- Provides consistent standard for naming, describing locating, managing, accessing and securing individual resources

- Provide interfaces and synchronization support for other system directories used by the company

- Organize information hierarchically to ease network use and management

- Flexibility to store a wide range of attributes in the directory and tightly control access to them at the attribute level (i.e. allow global access to a person object, but lock access to the Social Security Number attribute)

- Allow for creation of multiple copies of the directory and automatically replicate and synchronize changes among them

- Support multi-master replication for flexibility, high-availability, and performance

- Provide role-based delegation of administration rights to let administrators delegate specific administrative privileges and tasks to individual users and groups

- Provide highly flexible directory search and query functions

- Provide Internet-ready security services to protect data while facilitating access

- Support fully integrated public key infrastructure and Internet secure protocols (i.e. LDAP over SSL) to let organizations securely extend selected directory information beyond firewall perimeters

- Provide open synchronization mechanisms to ensure interoperability, replication and synchronization with multiple platforms

System Event Monitor

The key functions of this tool type should include:

- Ability to capture and log events

- Ability to modify event status

- Ability to validate event content

- Ability to correct event content errors

- Ability to generate systems events

- Ability to filter and threshold events, i.e. ability to filter events on the basis of time, number of occurrences, etc.

- Ability to initiate actions automatically

- Ability to execute pre-defined script based on date and time (for housekeeping)

- Ability, as the result of an event, to execute pre-defined scripts and test the success

- Ability to monitor managed entity (specific metrics)

- Correlation of incidents

- Ability to forward events to distributed event managers

- Ability to collect error statistics

- Ability to generate notifications in multiple formats

- Ability to maintain local log file on managed entity

- Ability to customize parameters as required

- Interface to Configuration Management Database

- Interface to Incident Management Database

- Ability to forward events to the Event Correlation Tool for analysis and action

- Real-time status display

- Ability to access event information locally and remotely

- Provide multiple interfaces such as Command Line, Batch or GUI

- Highly flexible capability for monitoring service levels and operational level targets

- Ability to configure monitoring/monitoring scripts to summarize data for reporting purposes

- Automatic alerts for service levels in jeopardy or missed service targets

- IT component downtime data capture and recording

- Capability to alert across different communication channels such as E-Mail, Pager, FAX, or PDA

- Ability to provide variety of notification messages, memos or letters using templates for common event situations

Test Data Generator

The key functions of this tool type should include:

- Ability to generate data records and populate data fields based on complex business logic

- Provide command line and GUI interfaces

- Flexible data record formatting functions

- Storage for multiple versions of test data

- Provision of listings of test data

- Ability to create volumes of unique records automatically without re-entry

- Capability to audit data generated to ensure unique key fields have been populated without replicated data

ITSM Logical Data Model

The following chart illustrates common data fields that should be found in each of the data repositories associated with ITSM.

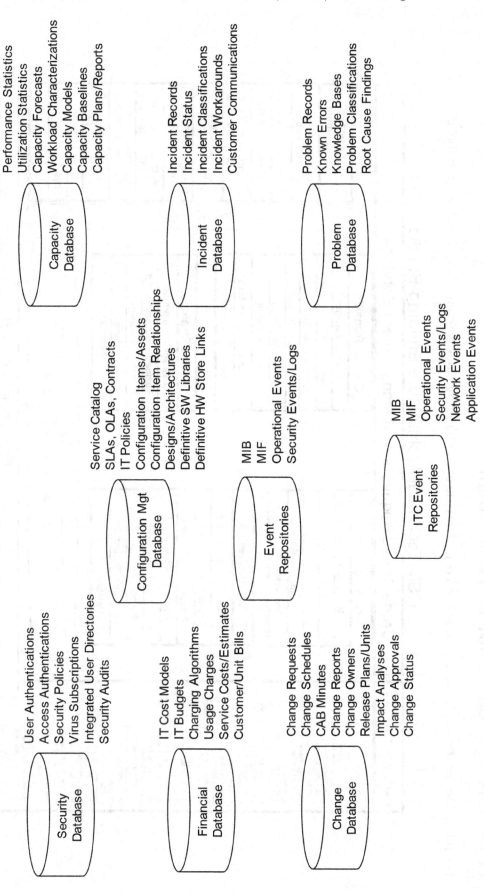

Security Database
- User Authentications
- Access Authentications
- Security Policies
- Virus Subscriptions
- Integrated User Directories
- Security Audits

Financial Database
- IT Cost Models
- IT Budgets
- Charging Algorithms
- Usage Charges
- Service Costs/Estimates
- Customer/Unit Bills

Change Database
- Change Requests
- Change Schedules
- CAB Minutes
- Change Reports
- Change Owners
- Release Plans/Units
- Impact Analyses
- Change Approvals
- Change Status

Capacity Database
- Performance Statistics
- Utilization Statistics
- Capacity Forecasts
- Workload Characterizations
- Capacity Models
- Capacity Baselines
- Capacity Plans/Reports

Configuration Mgt Database
- Service Catalog
- SLAs, OLAs, Contracts
- IT Policies
- Configuration Items/Assets
- Configuration Item Relationships
- Designs/Architectures
- Definitive SW Libraries
- Definitive HW Store Links

Event Repositories
- MIB
- MIF
- Operational Events
- Security Events/Logs

Incident Database
- Incident Records
- Incident Status
- Incident Classifications
- Incident Workarounds
- Customer Communications

Problem Database
- Problem Records
- Known Errors
- Knowledge Bases
- Problem Classifications
- Root Cause Findings

ITC Event Repositories
- MIB
- MIF
- Operational Events
- Security Events/Logs
- Network Events
- Application Events

ITSM Tooling Architecture Reference Model

The following chart provides an example of an overall architectural building block framework for ITSM support tools.

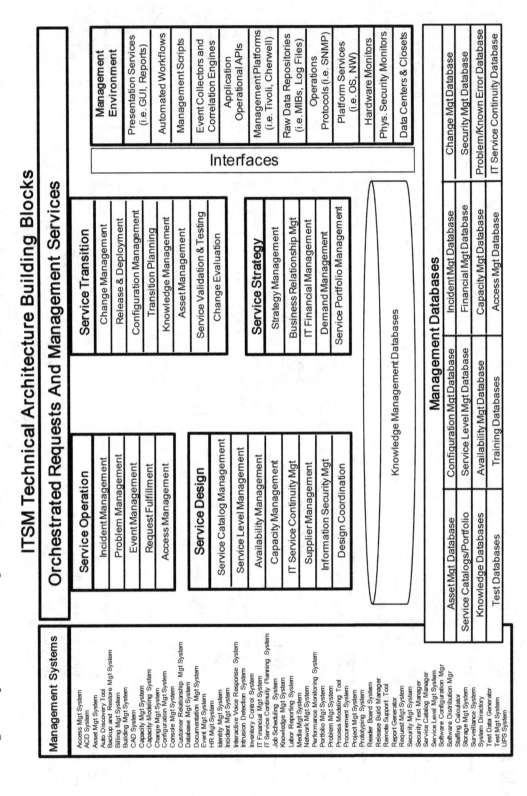

ITSM Technical Architecture Building Blocks

Chapter
13

ITSM Process Work Products

"If I had asked my customers what they wanted they would have said a faster horse."
—Henry Ford

Memo to get attention of Senior Management:

"Our IT support organization has recently discovered a new emerging industry approach that results in quick wins for increasing our company shareholder value, reduction in IT support costs and increases in customer satisfaction levels. We believe this can be put to work within our company by the end of this year. Is it worth 20 minutes to talk about this? We look forward to your thoughts."

The following pages identify typical Work Products that are produced by each ITSM process and function. Each table describes one ITSM process in terms of its Work Products. The associated columns indicate all other ITSM processes that will be impacted in some way by each Work Product. An **X** in any of these columns indicates that the Work Product is used by the ITSM process represented by that column.

This is not meant to be a definitive listing, but should provide a solid reference for understanding what key ITSM Work Products are produced by each process and which other processes use them.

Incident Management Work Products

Work Product	Service Desk	Incident	Problem	Change	Release	Config.	Service Level	Avail.	Capacity	IT Service Continuity	Financial	Security
Incident Status Reports	x	n/a	x		x	x	x	x	x			x
Incident Classification Schema	x	n/a	x			x	x					x
Incident Descriptions	x	n/a	x	x	x	x	x	x	x			x
Incident Workarounds	x	n/a	x	x	x	x	x	x	x	x		x
Incident Notification Lists	x	n/a	x	x	x	x	x	x	x	x		x
Recorded Incident Records	x	n/a	x	x	x	x		x	x			x
Incident Audit Reports		n/a	x			x	x	x				x
Incident Escalation Policy	x	n/a	x			x	x	x	x	x		x
Incident Historical Reports		n/a	x		x	x	x	x	x			x
Incident Process Metrics		n/a	x			x	x	x				
Incident Management Roles	x	n/a	x			x	x	x				x
Incident Responsibility Matrix	x	n/a	x			x	x	x				
Incident Job Descriptions	x	n/a	x			x	x	x			x	
Incident Tool Requirements	x	n/a	x				x	x	x		x	x
Incident Resolution RFCs	x	n/a	x	x		x	x	x	x			x
Closed Incidents	x	n/a	x			x	x	x				x
Incident Lifecycle Policy	x	n/a	x	x	x	x	x	x	x			x
Incident Procedures	x	n/a	x			x	x	x				x
Incident Prioritization Policy	x	n/a	x	x	x	x	x	x	x	x		x
Updated Incident Tickets	x	n/a	x	x	x	x	x	x	x			x
Incident Management Implementation Plan	x	n/a	x	x	x	x	x	x	x	x	x	
Incident Management Staffing Plan		n/a				x	x	x		x	x	
Infrastructure Incident Model	x	n/a	x		x	x	x	x	x	x		x

Problem Management Work Products

Work Product	Service Desk	Incident	Problem	Change	Release	Config.	Service Level	Avail.	Capacity	IT Service Continuity	Financial	Security
Problem Status Reports	X	X	n/a	X	X	X	X	X	X			X
Problem Classification Schema	X	X	n/a			X	X					X
Problem Descriptions	X	X	n/a	X	X	X	X	X	X			X
Known Error Descriptions	X	X	n/a	X	X	X	X	X	X			X
Recorded Problem Records	X	X	n/a	X	X	X		X	X			X
Problem Escalation Policy	X	X	n/a			X	X	X	X	X		X
Problem Historical Reports		X	n/a		X	X	X	X	X			X
Problem Process Metrics			n/a			X	X	X				
Problem Management Roles	X	X	n/a			X	X	X				
Problem Responsibility Matrix	X	X	n/a			X	X	X				X
Problem Job Descriptions		X	n/a			X	X	X			X	
Problem Tool Requirements		X	n/a				X	X	X		X	X
Error Resolution RFCs			n/a	X	X	X	X	X	X			X
Closed Problems	X	X	n/a			X	X	X				X
Problem Lifecycle Policy	X	X	n/a	X	X	X	X	X	X			X
Problem Procedures	X	X	n/a			X	X	X				X
Problem Prioritization Policy	X	X	n/a	X	X	X	X	X	X			X
Updated Problem Tickets	X	X	n/a	X	X	X	X	X	X			X
Problem/Known Error List	X	X	n/a		X	X	X	X	X			X
Problem Matching Policy	X	X	n/a			X		X				
Problem Analysis Toolkit			n/a			X		X				
Major Problem Review Meeting Minutes			n/a				X	X				X
Problem Management Implementation Plan	X		n/a	X	X	X	X	X	X	X	X	
Problem Management Staffing Plan			n/a			X	X	X			X	

Change Management Work Products

Work Product	Service Desk	Incident	Problem	Change	Release	Config.	Service Level	Avail.	Capacity	IT Service Continuity	Financial	Security
Change Status Reports	X		X	n/a	X	X	X	X	X	X		X
Change Categorization Schema		X		n/a		X	X	X				X
Change Descriptions	X	X	X	n/a	X	X	X	X	X	X		X
Recorded Change Records	X	X	X	n/a	X	X	X	X	X	X		X
Change Historical Reports		X	X	n/a	X	X	X	X	X	X		X
Change Process Metrics		X	X	n/a		X	X	X				
Change Management Roles	X	X	X	n/a	X	X	X	X	X			
Change Responsibility Matrix	X		X	n/a	X	X	X	X	X	X		X
Change Job Descriptions				n/a	X	X	X	X			X	
Change Tool Requirements				n/a	X		X	X		X	X	X
Closed Changes	X	X	X	n/a	X	X	X	X	X	X	X	X
Change Lifecycle Policy	X	X	X	n/a	X	X	X	X	X			X
Change Procedures	X	X	X	n/a	X	X	X	X	X	X		X
Change Prioritization Policy	X	X	X	n/a	X	X	X	X	X	X		X
Updated Change Tickets	X	X	X	n/a	X	X	X	X	X	X	X	X
PIR Meeting Minutes	X	X	X	n/a	X	X	X	X	X			X
Forward Schedule Of Changes	X	X	X	n/a	X	X	X	X	X	X		X
Requests For Change	X	X	X	n/a	X	X	X	X	X	X	X	X
CAB Meeting Minutes	X	X	X	n/a	X	X	X	X	X	X	X	X
CAB Policy	X	X	X	n/a	X	X	X	X	X	X	X	
PSA Notices	X	X	X	n/a	X	X	X	X	X	X		X
Change Impact Assessment Policy	X	X	X	n/a	X	X	X	X	X	X	X	X
Change Authorization Policy	X	X	X	n/a	X	X	X	X	X	X	X	X
Change Audit Results	X	X	X	n/a	X	X	X	X	X	X		

Work Product	Service Desk	Incident	Problem	Change	Release	Config.	Service Level	Avail.	Capacity	IT Service Continuity	Financial	Security
Standard Change Policy	×	×	×	n/a	×	×	×	×	×	×	×	×
Emergency Change Policy	×	×	×	n/a	×	×	×	×	×	×	×	×
CAB/EC Policy	×	×	×	n/a	×	×	×	×	×	×	×	×
Change Management Implementation Plan	×			n/a	×	×	×	×	×	×	×	
Change Management Staffing Plan				n/a		×	×	×			×	

Release Management Work Products

Work Product	Service Desk	Incident	Problem	Change	Release	Config.	Service Level	Avail.	Capacity	IT Service Continuity	Financial	Security
Release Status Reports	X	X	X	X	n/a	X	X	X	X	X		X
Release Process Metrics			X	X	n/a	X	X	X				
Release Management Roles	X	X	X	X	n/a	X	X	X	X			
Release Responsibility Matrix	X	X	X	X	n/a	X	X	X	X	X		X
Release Job Descriptions					n/a	X	X	X			X	
Release Tool Requirements					n/a		X	X			X	X
Release Lifecycle Policy	X	X	X	X	n/a	X	X	X	X	X		X
Release Procedures	X	X	X	X	n/a	X	X	X	X			X
Release Schedule	X	X	X	X	n/a	X	X	X	X	X	X	X
Release Policy	X	X	X	X	n/a	X	X	X	X	X		X
Release Plans and Descriptions	X	X	X	X	n/a	X	X	X	X	X	X	X
DHS Policy				X	n/a	X	X	X		X	X	X
DSL Policy				X	n/a	X	X	X		X	X	X
Release Test Plans			X	X	n/a	X	X	X	X	X		X
Release Rollout Plans	X	X	X	X	n/a	X	X	X	X	X	X	X
Release Back-out Plans		X	X	X	n/a	X	X	X	X	X		X
Release Communication Plans	X	X		X	n/a	X	X	X				X
Release Training Plans	X			X	n/a	X	X	X				X
Release Acceptance Certification Policy			X	X	n/a	X	X	X	X	X	X	X
Release Design Documents			X	X	n/a	X	X	X	X	X	X	X
Release Work Plans			X	X	n/a	X	X	X	X	X	X	X
Release Management Implementation Plan	X		X	X	n/a	X	X	X	X	X	X	
Release Management Staffing Plan					n/a	X	X	X			X	

Configuration Management Work Products

Work Product	Service Desk	Incident	Problem	Change	Release	Config.	Service Level	Avail.	Capacity	IT Service Continuity	Financial	Security
Configuration Status Reports	x	x	x	x	x	n/a	x	x	x	x	x	x
Configuration Relationship Reports	x	x	x	x	x	n/a	x	x	x	x		x
Recorded CI Records	x	x	x	x	x	n/a		x	x	x		x
Configuration Process Metrics			x	x	x	n/a	x	x	x	x	x	
Configuration Management Roles	x	x	x	x	x	n/a	x	x	x	x	x	
Configuration Responsibility Matrix	x	x	x	x	x	n/a	x	x	x	x	x	x
Configuration Job Descriptions				x	x	n/a	x	x		x	x	
Configuration Mgt. Tool Requirements	x	x	x	x	x	n/a	x	x	x	x	x	x
CI Database Schema	x	x	x	x	x	n/a	x	x	x	x		
CMDB Physical Design					x	n/a	x	x	x	x		x
Configuration Lifecycle Policy	x		x	x	x	n/a	x	x	x	x	x	x
Configuration Procedures	x	x	x	x	x	n/a	x	x	x	x	x	x
CI Naming Conventions				x	x	n/a	x	x				x
Updated CI Records	x		x	x	x	n/a	x	x	x		x	x
Configuration Audit Results	x	x	x	x	x	n/a	x	x	x	x	x	x
CI Scope Policy		x	x	x	x	n/a	x	x	x	x	x	x
CI Repository Sourcing Matrix				x	x	n/a	x	x	x	x	x	x
CMDB OLA Requirements						n/a	x	x	x	x	x	x
Configuration Control Policy				x	x	n/a	x	x			x	x
Configuration Baselines		x	x	x	x	n/a		x	x	x		x
Configuration Baseline Reports		x	x	x	x	n/a	x	x	x	x	x	x
License Management Control Policy	x	x	x	x	x	n/a	x	x	x	x	x	x
Configuration Mgt. Implementation Plan	x			x	x	n/a	x	x	x	x	x	
Configuration Management Staffing Plan						n/a	x	x			x	

Service Level Management Work Products

Work Product	Service Desk	Incident	Problem	Change	Release	Config.	Service Level	Avail.	Capacity	IT Service Continuity	Financial	Security
Service Catalog	x	x	x	x	x	x	n/a	x	x	x	x	x
Service Level Agreements	x	x	x	x	x	x	n/a	x	x	x	x	x
Operational Level Agreements	x	x	x	x	x	x	n/a	x	x	x	x	x
Underpinning Contracts	x	x	x	x	x	x	n/a	x	x	x	x	x
Service Level Management Process Metrics	x	x	x	x	x	x	n/a	x	x	x	x	
Service Level Management Roles	x	x	x	x	x	x	n/a	x	x	x	x	
Service Level Mgt. Responsibility Matrix	x	x	x	x	x	x	n/a	x	x	x	x	x
Service Level Management Job Descriptions						x	n/a	x			x	
Service Level Management Tool Requirements	x	x	x	x	x	x	n/a	x	x	x	x	x
Service Reports	x	x	x	x	x	x	n/a	x	x	x	x	x
Vendor Reports	x	x	x	x	x	x	n/a	x	x	x	x	x
Service Monitoring Strategy	x	x	x		x	x	n/a	x	x	x	x	x
SIP Policy				x	x	x	n/a	x	x	x	x	x
Service Communication Plan	x			x	x	x	n/a	x	x	x	x	x
SLA Structure	x	x	x	x	x	x	n/a	x	x	x	x	x
SLA Rollout Strategy	x	x	x	x	x	x	n/a	x	x	x	x	x
Service Review Meeting Agenda				x	x	x	n/a	x		x		
Service Escalation Policy	x	x	x	x	x	x	n/a	x	x	x	x	x
Service Level Mgt. Implementation Plan	x		x	x		x	n/a	x	x	x	x	
Service Level Management Staffing Plan						x	n/a	x		x	x	

Availability Management Work Products

Work Product	Service Desk	Incident	Problem	Change	Release	Config.	Service Level	Avail.	Capacity	IT Service Continuity	Financial	Security
Service Breakdown Matrix			x	x	x	x	x	n/a	x	x	x	x
Monitoring Analysis Matrix		x	x	x	x	x		n/a	x		x	x
CI Organizational Matrix	x	x	x	x	x	x	x	n/a	x	x	x	x
Availability Requirements Inventory			x	x	x	x	x	n/a	x	x		x
Availability Reports	x	x	x			x	x	n/a	x	x		x
Availability Management Process Metrics	x	x	x	x	x	x	x	n/a	x	x	x	
Availability Management Roles	x	x	x	x	x	x	x	n/a	x	x	x	
Availability Mgt. Responsibility Matrix	x	x	x	x	x	x	x	n/a	x	x	x	x
Availability Management Job Descriptions						x	x	n/a			x	
Availability Management Tool Requirements	x	x	x	x	x	x	x	n/a	x	x	x	x
Availability Issues Log	x	x	x	x	x	x	x	n/a	x	x	x	x
Availability Recommendations Inventory	x	x	x	x	x	x	x	n/a	x	x	x	x
Availability RFCs		x	x	x	x	x	x	n/a	x	x	x	x
Availability Plan	x	x	x	x	x	x	x	n/a	x	x	x	x
Availability Policy	x	x	x	x	x	x	x	n/a	x	x	x	x
Service Level Assessments						x	x	n/a	x	x	x	x
Infrastructure Maintenance Plan	x	x	x	x	x	x	x	n/a	x	x	x	x
Infrastructure Risk Analyses						x	x	n/a	x	x	x	x
Availability Management Implementation Plan	x			x	x	x	x	n/a	x	x	x	
Availability Management Staffing Plan						x	x	n/a			x	

Capacity Management Work Products

Work Product	Service Desk	Incident	Problem	Change	Release	Config.	Service Level	Avail.	Capacity	IT Service Continuity	Financial	Security
Capacity Management Policy	x	x	x	x	x	x	x	x	n/a	x	x	x
Capacity Management Process Metrics	x	x	x	x	x	x	x	x	n/a	x	x	
Capacity Management Roles	x	x	x	x	x	x	x	x	n/a	x	x	
Capacity Management Responsibility Matrix	x	x	x	x	x	x	x	x	n/a	x	x	x
Capacity Management Job Descriptions						x	x	x	n/a		x	
Capacity Management Tool Requirements	x	x	x	x	x	x	x	x	n/a	x	x	x
Capacity Plan	x	x	x	x	x	x	x	x	n/a	x	x	x
Capacity Reports	x	x	x	x	x	x	x	x	n/a	x	x	x
Performance Reports	x	x	x	x	x	x	x	x	n/a	x	x	x
Business Forecast Inventory	x	x	x				x	x	n/a	x	x	
Workload Characterizations			x			x		x	n/a	x		
Capacity Baselines						x		x	n/a	x		
Capacity Requirements		x	x	x	x	x	x	x	n/a	x	x	x
Capacity Models				x	x	x	x	x	n/a		x	
Capacity Thresholds	x	x	x		x	x	x	x	n/a	x		x
Capacity Issues Log	x	x	x			x	x	x	n/a	x	x	x
Capacity and Tuning Recommendations List			x	x	x	x	x	x	n/a	x	x	x
Service Level Assessments					x	x	x	x	n/a		x	x
Capacity Maintenance Plan	x	x	x	x	x	x	x	x	n/a	x	x	
Capacity RFCs	x	x	x	x	x	x	x	x	n/a	x	x	x
Application Sizing Estimates				x	x	x	x	x	n/a	x	x	x
Capacity Monitoring Matrix	x	x	x		x	x	x	x	n/a	x	x	x
Capacity Management Implementation Plan	x			x		x	x	x	n/a	x	x	
Capacity Management Staffing Plan						x	x	x	n/a		x	

IT Service Continuity Management (ITSCM) Work Products

Work Product	Service Desk	Incident	Problem	Change	Release	Config.	Service Level	Avail.	Capacity	IT Service Continuity	Financial	Security
ITSCM Policy	x	x	x	x	x	x	x	x	x	n/a	x	x
VBF Inventory	x	x	x	x	x	x	x	x	x	n/a	x	x
ITSCM Process Metrics	x	x	x	x	x	x	x	x	x	n/a	x	
ITSCM Roles	x	x	x	x	x	x	x	x	x	n/a	x	
ITSCM Responsibility Matrix	x	x	x	x	x	x	x	x	x	n/a	x	x
ITSCM Job Descriptions						x	x	x		n/a	x	
ITSCM Tool Requirements	x	x	x	x	x	x	x	x	x	n/a	x	x
IT Impact Analysis	x	x	x	x	x	x	x	x	x	n/a	x	x
ITSCM Risk Assessment						x	x	x		n/a	x	x
ITSCM Strategy	x	x	x	x	x	x	x	x	x	n/a	x	x
IT Service Continuity Plan	x	x	x	x	x	x	x	x	x	n/a	x	x
ITSCM RFCs	x	x	x	x	x	x	x	x	x	n/a	x	x
ITSCM Test Plan	x	x	x	x	x	x	x	x	x	n/a	x	x
ITSCM Test Results	x	x	x	x	x	x	x	x	x	n/a	x	x
ITSCM Communications and Training Plan	x					x	x	x		n/a	x	x
ITSCM Audit Results						x	x	x		n/a	x	x
Assurance Signoffs						x	x	x		n/a		x
Invocation Policy	x	x	x	x	x	x	x	x	x	n/a	x	x
ITSCM Implementation Plan	x	x	x	x	x	x	x	x	x	n/a	x	
ITSCM Staffing Plan						x	x	x	x	n/a	x	

IT Financial Management Work Products

Work Product	Service Desk	Incident	Problem	Change	Release	Config.	Service Level	Avail.	Capacity	IT Service Continuity	Financial	Security
Financial Management Policy	×	×	×	×	×	×	×	×	×	×	n/a	×
Financial Management Process Metrics	×	×	×	×	×	×	×	×	×	×	n/a	
Financial Management Roles	×	×	×	×	×	×	×	×	×	×	n/a	
Financial Management Responsibility Matrix	×	×	×	×	×	×	×	×	×	×	n/a	×
Financial Management Job Descriptions						×	×	×			n/a	
Financial Management Tool Requirements				×	×	×	×	×			n/a	×
Cost Estimates	×	×	×	×	×	×	×	×	×	×	n/a	×
IT Charging Policy	×					×	×				n/a	
IT Budget Policy				×	×	×	×	×	×		n/a	
IT Budget Model	×	×	×	×	×	×	×	×	×	×	n/a	
IT Budget and Accounting Reports							×	×	×	×	n/a	
IT Charging Reports	×						×	×	×	×	n/a	×
IT Charging Bills	×						×				n/a	

Service Desk Work Products

Work Product	Service Desk	Incident	Problem	Change	Release	Config.	Service Level	Avail.	Capacity	IT Service Continuity	Financial	Security
Service Desk Reports	n/a	x	x			x	x	x			x	x
Service Desk Functional Metrics	n/a	x	x			x	x	x				
Service Desk Roles	n/a	x	x			x	x	x				
Service Desk Responsibility Matrix	n/a	x	x			x	x	x				x
Service Desk Job Descriptions	n/a	x	x			x	x	x			x	
Service Desk Tool Requirements	n/a	x	x				x	x	x		x	x
Service Desk Staffing Plan	n/a	x				x	x	x			x	
Service Desk Implementation Plan	n/a	x		x	x	x	x	x			x	
Service Desk Communication Policy	n/a	x	x	x	x	x	x	x	x	x	x	x
Service Desk Awareness Campaign	n/a	x	x	x	x	x	x	x	x	x	x	
Service Desk Security Policy	n/a	x	x	x	x	x	x	x	x	x	x	x
Skills Maintenance Plan	n/a	x	x	x	x	x	x	x	x	x	x	
Service Desk Logical Design	n/a	x	x	x	x	x	x	x	x	x	x	x
Service Desk Physical Design	n/a	x	x	x	x	x	x	x	x	x	x	x
User Survey Policy	n/a	x	x	x	x	x	x	x	x	x	x	
Customer Database Policy	n/a	x	x	x	x	x	x	x		x		x
Service Desk Archiving Policy	n/a	x	x	x	x	x	x	x	x	x	x	x

Security Management Work Products

Work Product	Service Desk	Incident	Problem	Change	Release	Config.	Service Level	Avail.	Capacity	IT Service Continuity	Financial	Security
Security Policy	x	x	x	x	x	x	x	x	x	x	x	n/a
Asset Classification and Control Strategy	x	x	x	x	x	x	x	x	x	x	x	n/a
Access Control Strategy	x	x	x	x	x	x	x	x	x	x	x	n/a
Security Escalation Policy	x	x	x	x	x	x	x	x	x	x	x	n/a
Inventory of Legal Compliance Requirements					x	x	x	x	x	x		n/a
Security Communications Strategy	x			x			x			x		n/a
Security Monitoring Plan	x	x	x				x	x				n/a
Security Audit Reports							x	x				n/a
Security Monitoring Reports							x	x				n/a
Security Management Function Metrics							x					n/a
Security Management Roles							x					n/a
Security Management Responsibility Matrix	x	x	x	x	x	x	x	x	x	x	x	n/a
Security Management Job Descriptions							x	x				n/a
Security Management Tool Requirements					x	x	x	x	x	x		n/a
Security Issues Log	x	x	x				x	x	x	x		n/a
Security Recommendations Inventory				x	x	x	x	x		x		n/a
Security RFCs	x	x	x	x	x	x	x	x		x		n/a
Security Maintenance Plan				x	x		x	x		x		n/a
Security Risk Analyses							x	x		x		n/a
Security Management Implementation Plan	x			x	x	x	x	x	x	x	x	n/a
Security Management Staffing Plan						x	x	x		x		n/a

Chapter
14

ITSM Operating Roles

> Adopting an IT Service Management culture is like cultivating
> a garden. The seeds and flowers you grow today will shrivel
> and die over time if not given ongoing attention.

The following pages identify typical roles for operating each ITSM process and the ITSM infrastructure. Note that a role defines a discreet set of tasks to be performed. It is not equivalent to a job description.

Job descriptions may be made up of one or more of the roles described here. In some cases, a role may split organizationally into multiple jobs or departments.

The pages that follow list common roles for each ITSM process, role descriptions and key activities associated with them.

Incident Management Roles

Overview
:   The key roles involved with executing this process are summarized as follows:
    - Incident Management Process Owner
    - Incident Manager
    - Incident Analyst (First Line)
    - Incident Analyst (Second Line)
    - Incident Auditor
    - Subject Matter Expert

Incident Management Process Owner
:   Primary role is to ensure executive support of the Incident Management process, co-ordinate the various functions and work activities at all levels of the process, provide the authority or ability to make changes in the process as required, and manage the process end-to-end so as to ensure optimal overall performance. Key activities include:

    - Communicate the organizational vision and the process's strategic goals to business units and IT organizations
    - Identify and assimilate similar and/or overlapping activities/ initiatives within IT
    - Coordinate inter-process changes with other process owners
    - Provide process ownership through design, implementation and continuous improvement activities in the process life cycle
    - Work with all functions within the IT community to ensure processes are executed as designed and measured accurately and completely
    - Ensure alignment of the process to the corporate and IT strategy
    - Create a new environment by displaying a motivational impatience for results (be a behavior change leader)
    - Publicize activities and build commitment and consensus around Incident Management activities
    - Facilitate resolution of interface problems with other process owners
    - Communicate with and manage the expectations of customers of IT Services

Incident Manager	Responsible for managing Incident staff and oversight of the Incident Management process for IT. Key activities include:

- Driving the efficiency and effectiveness of the Incident Management process
- Producing Incident Management reports and information
- Managing the work of Incident support staff (first-and second-line)
- Monitoring the effectiveness of Incident Management and making recommendations for improvement
- Developing and maintaining Incident Management systems
- Establishment and execution of ongoing audit activities within Incident Management

Incident Analyst (First Line)	Supports the Incident Management process by providing an operational single point of contact to manage incidents to resolution. Key activities include:

- Incident registration
- Routing service requests to support groups when Incidents are not closed
- Initial support and analyze for correct classification
- Ownership, monitoring, tracking and communication
- Resolution and recovery of Incidents not assigned to second-line support
- Closure of Incidents
- Monitoring the status and progress towards resolution of assigned Incidents
- Keeping help desk informed about incident progress
- Escalating the process as necessary per established escalation policies

Incident Analyst (Second Line)	Provides a specialist skill or support role to receive Incidents that cannot be resolved by First Line Incident Analysts to investigate and coordinate resolution and recovery actions for assigned Incidents Internal to IT. Key activities include:

- Handling service requests
- Monitoring Incident details, including the Configuration Items affected
- Incident investigation and diagnosis (including resolution where possible)
- Detection of possible Problems and the assignment of them to the Problem Management team for them to raise Problem records
- The resolution and recovery of assigned Incidents
- Participation in audit activities related to the Incident Management process
- Monitoring the status and progress towards resolution of assigned Incidents
- Keeping affected business partners informed about progress
- Escalating the process as necessary per established escalation policies

Incident Auditor	Responsible for auditing recorded incident records to ensure that incidents have been accurately classified and prioritized. Key activities include:

- Understanding current incident classification and prioritization criteria
- Conducting periodic and regular audits of incident tickets
- Identifying audit failures to Incident Management staff
- Producing audit reports on Incident quality

Subject Matter Expert	Assists with incident analysis and resolution activities within area of expertise. Key activities include:

- Analyzes incidents to identify service restoration actions to be taken
- Assists Incident Management staff with classification and prioritization of incidents
- Assists Incident Management staff with identifying the impact of incidents
- Takes incident resolution actions to restore service to customers

Problem Management Roles

Overview	The key roles involved with executing this process are summarized as follows:

- Problem Management Process Owner
- Problem Manager
- Problem Owner
- Subject Matter Expert

Problem Management Process Owner	Primary role is to ensure executive support of the Problem Management process, co-ordinate the various functions and work activities at all levels of the process, provide the authority or ability to make changes in the process as required, and manage the process end-to-end so as to ensure optimal overall performance. Key activities include:

- Communicate the organization vision and the process's strategic goals to business units and IT organizations
- Identify and assimilate similar and/or overlapping activities/initiatives within IT
- Co-ordinate inter-process changes with other process owners
- Provide process ownership through design, implementation and continuous improvement activities in the process life cycle
- Work with all functions within the IT community to ensure processes are executed as designed and measured accurately and completely
- Ensure alignment of the process to the corporate and IT strategy
- Create a new environment by displaying a motivational impatience for results (be a behavior change leader)
- Publicize activities and build commitment and consensus around Problem Management activities
- Facilitate resolution of interface problems with other process owners
- Communicate with and manage the expectations of customers of Problem Management

Problem Manager	Responsible for reviewing problem trends and proactively taking actions to identify problems and remove errors for a department or business unit. Key activities include:

- Reviews the efficiency and effectiveness of the Problem control process
- Produces Problem Management reports and management information
- Manages Problem support staff
- Allocates resources for the support effort
- Monitors the effectiveness of error control and makes recommendations for improvements
- Develops and maintains Problem and error control systems
- Reviews the efficiency and effectiveness of proactive Problem Management activities
- Identifies trends and potential Problem sources (by reviewing Incident and Problem analyses)
- Prevents the replication of Problems across multiple systems

Problem Owner	Provides a single point of contact for one or more problems and is responsible for ownership and coordination of actions of those problems to analyze for root cause, identify Known Error and coordinating actions to remove the error. Key activities include:

- Reviews Incident data to analyze assigned problems
- Investigates assigned problems through to resolution or error identification
- Coordinates actions of others as necessary to assist with analysis and resolution actions for problems and Known Errors
- Raises RFCs to clear errors
- Monitors progress on the resolution of Known Errors and advises Incident Management staff on the best available Work-Around for Incidents related to unresolved Problems/Known Errors
- Assists with the handling of major Incidents and identifying the root causes

Subject Matter Expert	Assists with problem analysis and resolution activities within area of expertise. Key activities include:

- Analyzes problems to identify Root Cause
- Identifies problem work-around
- Takes problem resolution actions to remove errors from the infrastructure
- Assists Problem Owner in investigating assigned problems through to resolution or error identification

Change Management Roles

Overview	The key roles involved with executing this process are summarized as follows:

- Change Management Process Owner
- Change Manager
- Change Analyst
- Change Owner
- CAB Member
- Subject Matter Expert

Change Management Process Owner

Primary role is to ensure executive support of the Change Management process, co-ordinate the various functions and work activities at all levels of the process, provide the authority or ability to make changes in the process as required, and manage the process end-to-end so as to ensure optimal overall performance. Key activities include:

- Communicate the organization vision and the process's strategic goals to business units and IT organizations
- Identify and assimilate similar and/or overlapping activities/ initiatives within IT
- Co-ordinate inter-process changes with other process owners
- Provide process ownership through design, implementation and continuous improvement activities in the process life cycle
- Work with all functions within the IT community to ensure processes are executed as designed and measured accurately and completely
- Ensure alignment of the process to the corporate and IT strategy
- Create a new environment by displaying a motivational impatience for results (be a behavior change leader)
- Publicize activities and build commitment and consensus around Change Management activities
- Facilitate resolution of interface problems with other process owners
- Communicate with and manage the expectations of customers of Change Management

Change Manager	Represents IT or an IT department or group and is responsible for all changes within that area. Key activities include:

- Receive, log and allocate a priority, in collaboration with the initiator, to all RFCs Reject any RFCs that are totally impractical
- Table all RFCs for a CAB meeting, issue an agenda and circulate all RFCs to CAB members in advance of meetings to allow prior consideration
- Decide which people will come to which meetings, who get specific RFCs depending on the nature of the RFC, what is to be changed, and people's areas of expertise
- Convene urgent CAB or CAB/EC meetings for all urgent RFCs
- Chair all CAB and CAB/EC meetings
- After consideration of the advice given by the CAB or CAB/EC, authorize acceptable Changes
- Issues the Forward Schedule of Changes (FSC)
- Liaise with all necessary parties to coordinate Change building, testing and implementation, in accordance with schedules
- Update the Change log with all progress that occurs, including any actions to correct problems and/or to take opportunities to improve service quality
- Review all implemented Changes to ensure that they have met their objectives Refer back any that have been backed out or have failed
- Review all outstanding RFCs awaiting consideration or awaiting action
- Analyze Change records to determine any trends or apparent problems that occur Seek rectification with relevant parties
- Close RFCs
- Produce regular and accurate management reports

Change Analyst Administrative support role to handle tasks that log changes, administer change management tools, assemble and produce periodic change reports and any other tasks that assist the Change Management process. Key activities include:

- Administers Change Management support tools
- Assembles and produces periodic Change Management reports
- Handles ad-hoc requests for Change Management status or one-time information retrieval/reports
- Handles general communications throughout workflow as needed to ensure change packages are assembled properly and communicated
- Manages Change Management distribution lists

Change Owner Single point of contact for requesting a change—submits requests for changes (RFCs) and assembles submission packages and represents business unit or area interests regarding the change being requested. This role starts with the RFC submission to Change management and ends once the Change has been successfully implemented or permanently rejected. Key activities include:

- Fills out RFC form
- Assembles submission package for Change
- Ensures requirements for Change have been provided and provides documentary proof
- Single point of contact to the Change Management process to represent the request of the change
- Responds to issues regarding coordination and approval of the change as necessary
- Responsible for verifying that the change has been implemented successfully

CAB Member	Represents a business Unit, department, customers or other area within the business organization or IT area that will be impacted by a change. Key activities include:

- Reviews all submitted RFCs as appropriate and provide details of their likely impact, the implementation resources, and the ongoing costs of all Changes
- Attend all relevant CAB or CAB/EC meetings Consider all Changes on the agenda and give an opinion on which Changes should be authorized Participate in the scheduling of all Changes
- (CAB/EC only) Be available for consultation should an urgent Change be required
- Provide advice to Change Management on aspects of proposed urgent Changes

Subject Matter Expert	Assists with analysis of changes within area of expertise. Key activities include:

- Analyzes changes for applicability and impact
- May assist CAB with decision making input for changes
- May be asked to document Risk Memo for an escalated change

Release Management Roles

Overview	The key roles involved with executing this process are summarized as follows:

- Release Management Process Owner
- Release Manager
- Test Manager
- Release Owner
- Subject Matter Expert

Release Management Process Owner	Primary role is to ensure executive support of the Release Management process, coordinate the various functions and work activities at all levels of the process, provide the authority or ability to promote releases, as required, and manage the process end-to-end so as to ensure optimal overall performance. Key activities for this role include:

- Communicate the organization vision and the process's strategic goals to business units and IT organizations
- Identify and assimilate similar and/or overlapping activities/ initiatives within IT
- Co-ordinate inter-process releases with other process owners
- Provide process ownership through design, implementation and continuous improvement activities in the process life cycle
- Work with all functions within the IT community to ensure processes are executed as designed and measured accurately and completely
- Ensure alignment of the process to the corporate and IT strategy
- Create a new environment by displaying a motivational impatience for results (be a behavior change leader)
- Publicize activities and build commitment and consensus around Release Management activities
- Facilitate resolution of interface problems with other process owners
- Communicate with and manage the expectations of customers of Release Management
- Report process KPIs
- Provide process management reports
- Conduct periodic process audits

Release Manager	Represents a business, IT development project or IT unit and is responsible for all releases within that area. Key activities include:

- Develop release strategy
- Publish a build plan
- Publish a testing plan
- Publish deployment plan
- Conduct release plan reviews
- Receive, log, qualify and assign all release requests
- Participate in release control gates (as needed)
- Develop installation scripts
- Develop configuration scripts
- Develop roll back procedures
- Develop release procedures
- Develop test scripts
- Develop test readiness review
- Develop bill of materials
- Procure release components
- Works closely with the Test Manager to:
  - Review releases and assign appropriate release testing tasks
  - Compile and Review the Testing Deliverables
  - Conduct testing as needed
  - Conduct supporting documentation review
- Schedule release readiness reviews
- Conduct release readiness reviews
- Disposition releases
- Communicate release dispositions
- Participate in CAB meeting
- Deploy release
- Communicate release status
- Update Master Release Calendar & FSC
- Secure software in DSL
- Decommission Configuration Items (CIs)
- Participate in Post Implementation Review (PIR)
- Establish security permissions for DSL and access to managed environments
- Champion the new processes and encourage project managers and application teams to use them
- Report exceptions to the Release process
- Automate Release procedures wherever possible
- Ensure that all Customer, support staff and Service Desk staff are trained and provided with the appropriate documentation and information

Test Manager	Primary role is to ensure that proper testing occurs for all releases into the managed environments. Key activities include:

- Work closely with the Release Manager
- Review releases and assign appropriate release testing tasks
- Compiles and Review the Testing Deliverables
- Conduct installation procedure tests
- Conduct functional testing
- Conduct performance testing
- Conduct integration testing
- Conduct user acceptance testing
- Conduct operational readiness testing
- Conduct back out testing
- Conduct supporting documentation review
- Compile test results
- Conduct release test review
- Conduct post release testing

Release Owner	Single point of contact for the release being requested Represents business unit or area interests regarding the release being requested This role starts with the release submission to Release Management and ends once the release has been successfully implemented. Key activities include:

- Assemble submission package for a release
- Ensure requirements for a release have been provided and provides documentary proof
- Act as single point of contact to the Release Management process to represent the request for release
- Respond to issues regarding coordination of the release as necessary
- Hold responsibility for verifying that the release has been implemented successfully

Subject Matter Expert	Assists with analysis of releases within area of expertise. Key activities include:

- Analyze releases for applicability and impact
- Assist with release implementation and testing activities as needed
- Assist with integration and migration of releases to production status

Configuration Management Roles

Overview	The key roles involved with executing this process are summarized as follows:

- Configuration Management Process Owner
- Configuration Manager
- Configuration Analyst
- Configuration Librarian

Configuration Management Process Owner

Primary role is to ensure executive support of the Configuration Management process, co-ordinate the various functions and work activities at all levels of the process, provide the authority or ability to make changes in the process as required, and manage the process end-to-end so as to ensure optimal overall performance. Key activities include:

- Communicate the organization vision and the process's strategic goals to business units and IT organizations
- Identify and assimilate similar and/or overlapping activities/ initiatives
- Co-ordinate inter-process changes with other process owners
- Provide process ownership through design, implementation and continuous improvement activities in the process life cycle
- Work with all functions within the IT community to ensure processes are executed as designed and measured accurately and completely
- Ensure alignment of the process to the corporate and IT strategy
- Create a new environment by displaying a motivational impatience for results (be a behavior change leader)
- Publicize activities and build commitment and consensus around Configuration Management activities
- Facilitate resolution of interface problems with other process owners
- Communicate with and manage the expectations of customers of the Configuration Management process

Configuration Manager	Is responsible for all Configuration Management activities within IT. Key activities include:

- Plan for Configuration Management databases and activities
- Identify Configuration Items
- Control Configuration Item information
- Perform status accounting
- Perform verification and audit of Configuration Management databases
- Provide management information about Configuration Management quality and operations

Configuration Analyst	Provides facilities for identifying, associating names, controlling, collecting data from, the continuous operation of and the termination of connections of CIs within a management domain. Key activities include:

- Setting and resetting of parameters for the routine maintenance of CIs
- Associating names with CIs and sets of CIs within the defined schema
- Initializing, resetting and closing down CIs
- Collecting information about the current state of CIs
- Receiving and obtaining announcements of significant changes in the state or condition of CIs
- Changing the configuration of CIs or sets of CIs

Configuration Librarian	Responsible for maintaining up-to-date (and historical) records of configuration items. Key activities include:

- Maintain the Configuration Management Database
- Respond to requests for CI changes and updates from Change Management
- Provide CI information upon request

Service Level Management Roles

| Overview | The key roles involved with executing this process are summarized as follows: |

- Service Level Management Process Owner
- Service Owner
- Service Analyst
- Subject Matter Expert
- Business Unit Service Representative

Service Level
Management
Process Owner

Primary role is to ensure executive support of the Service Level Management process, co-ordinate the various functions and work activities at all levels of the process, provide the authority or ability to make changes in the process as required, and manage the process end-to-end so as to ensure optimal overall performance. Key activities include:

- Communicate the organization vision and the process's strategic goals to business units and IT organizations
- Identify and assimilate similar and/or overlapping activities/ initiatives within IT
- Co-ordinate inter-process changes with other process owners
- Provide process ownership through design, implementation and continuous improvement activities in the process life cycle
- Work with all functions within the IT community to ensure processes are executed as designed and measured accurately and completely
- Ensure alignment of the process to the corporate and IT strategy
- Create a new environment by displaying a motivational impatience for results (be a behavior change leader)
- Publicize activities and build commitment and consensus around Service Level Management activities
- Facilitate resolution of interface problems with other process owners
- Communicate with and manage the expectations of customers of IT Services

Service Owner   Represents a business unit and/or department and is responsible for all Service Level Management activities within that area. Key activities include:

- Create and maintain a catalog of existing services offered by the local service delivery organization
- Formulate, agree and maintain an appropriate SLM structure for the local Service Delivery
- Negotiate, agree and maintain SLAs in conjunction with business units
- Negotiate, agree and maintain OLAs within the local Service Delivery Center
- Ensure appropriate OLAs/SLAs in place to support any new services
- Analyze and review actual service performance against SLAs and OLAs
- Provide regular reports on service performance and achievement to the Service Manager
- Organize and maintain a regular Service Level review process with both key business and IT representatives
- Review SLA targets and metrics where necessary
- Review OLA targets and metrics where necessary
- Review third party underpinning agreements where necessary
- Agree appropriate actions to maintain or improve service levels
- Initiate and coordinate actions to maintain or improve service levels
- Act as a coordination point for any temporary changes to service levels

Service Analyst   Administrative support role to handle tasks that gather service data, administer service reporting tools, administer service catalog maintenance, assemble and produce service reports and any other tasks that assist the Service Level Management process. Key activities include:

- Administer Service Level Management support tools
- Assemble and produces Service Level Management reports
- Handle ad-hoc requests for Service Level Management status or one-time information retrieval/reports
- Manage service reporting communication distribution lists
- Maintain the Service Catalog
- Gather service data as needed for reporting and communications

Subject Matter Expert	Assists with service analysis and support activities within area of expertise. Key activities include:

- Interface between requests for services and IT support groups that provide services
- Provide expertise specific to a business function
- Provide expertise specific to negotiations and contracting with outside supplier vendors
- Assist Service Manager in investigating assigned service issues through to resolution

Business Unit Service Representative	Single point of contact for one or more business units to represent IT services. Key activities include:

- Act as a single point of contact for one or more Business Units
- Identify service needs for the Business Units represented to IT
- Escalate Business Unit(s) service issues to the Service Manager
- Communicate service status on service issues to the Business Unit(s)
- Assist in SLA negotiation efforts with the Business Unit(s)
- Report on quality of services rendered to the Business Unit(s)

Availability Management Roles

Overview	The key roles involved with executing this process are summarized as follows: • Availability Management Process Owner • Availability Manager • Availability Analyst • Subject Matter Expert • Availability Architect

Availability Management Process Owner	Primary role is to maintain the availability and reliability of IT services to ensure that IT can effectively meet service targets in accordance with planned business objectives. Key activities include: • Define and deploy the Availability Management process for execution and compliance across IT support organizations • Co-ordinate inter-process changes with other process owners • Provide process ownership through design, implementation and continuous improvement activities in the process life cycle • Work with all functions within the IT community to ensure processes are executed as designed and measured accurately and completely • Ensure alignment of the process to the corporate and IT strategy • Create a new environment by displaying a motivational impatience for results (be a behavior change leader) • Publicize activities and build commitment and consensus around Availability Management activities • Facilitate resolution of interface problems with other process owners • Ensure that the Availability Management process, its associated techniques and methods are regularly reviewed and audited, and that all of these are subjected to continuous improvement and remain fit for purpose

Availability Manager	This role has prime responsibility for maintaining the availability and reliability of IT services to ensure that IT can effectively meet service targets in accordance with planned business objectives. Key activities include:

- Optimize the availability of the IT infrastructure to deliver cost effective improvements that deliver tangible benefits to business units and customers
- Provide holistic management of availability that includes people and processes as well as technology
- Take actions to achieve reductions in frequency and duration of incidents that impact IT availability
- Ensure shortfalls in IT availability are recognized and appropriate corrective actions are identified and progressed
- Create and maintain a forward looking availability plan aimed at improving the overall availability of IT services and infrastructure components to ensure that existing and future availability requirements can be met
- Provide regular reports on availability to the Service Manager
- Organize and maintain a regular availability review process with both key business and IT representatives
- Agree appropriate actions to maintain or improve availability levels
- Initiate and coordinate actions required to maintain or improve availability across IT units
- Act as a coordination point for changes to availability levels when needed

Availability Analyst	Primary role is to analyze existing availability issues and problems to determine ways to improve availability at acceptable cost levels and determine availability requirements for new IT solutions and service changes. Key activities include:

- Provide a range of IT availability reporting to ensure that agreed levels of availability, reliability and maintainability are measured and monitored on an ongoing basis
- Determine the availability requirements from the business for new or enhanced IT services
- Establish measures and reporting that reflect business, user and IT support organization requirements
- Monitor actual availability achieved versus targets and to ensure shortfalls are addressed
- Participate in Change Control meetings to assess and authorize changes from an availability perspective
- Conduct availability risk assessment for existing services
- Assist in SLA negotiation efforts from an availability capability standpoint
- Define the key targets of availability required for the IT infrastructure and its components that underpin a new or enhanced IT service as the basis for an SLA agreement
- Analyze and review actual availability levels achieved against SLAs and OLAs and UCs
- Maintain an awareness of technology advancements and best practices that support availability
- Gather availability data as needed for reporting and communications

Subject Matter Expert	Assists with analysis of availability issues, strategies and activities within area of expertise. Key activities include:

- Interface between Availability Management initiatives and IT support groups that underpin services
- Provide expertise specific to a business or technology function
- Provide expertise specific to hardware/software solutions and vendor specific solutions
- Assist Availability Manager in investigating assigned availability issues through to resolution

Availability Architect	Assists Availability Management initiatives by providing overall strategy and design direction. Key activities include:

- Create availability and recovery design criteria to be applied to new or enhanced infrastructure design
- Ensure IT services are designed to deliver the required levels of availability required by business units and customers
- Ensure the levels of IT availability required are cost justified
- Identify availability needs for the Business Units represented to IT
- Document availability blueprints and designs as needed

Capacity Management Roles

Overview	The key roles involved with executing this process are summarized as follows: • Capacity Management Process Owner • Capacity Manager • Capacity Analyst • Subject Matter Expert • Capacity Architect
Capacity Management Process Owner	Primary role is to maintain the availability and reliability of IT services to ensure that IT can effectively meet service targets in accordance with planned business objectives. Key activities include: • Define and deploy the Capacity Management process for execution and compliance across IT support organizations • Co-ordinate inter-process changes with other process owners • Provide process ownership through design, implementation and continuous improvement activities in the process life cycle • Work with all functions within the IT community to ensure processes are executed as designed and measured accurately and completely • Ensure alignment of the process to the corporate and IT strategy • Create a new environment by displaying a motivational impatience for results (be a behavior change leader) • Publicize activities and build commitment and consensus around Capacity Management activities • Facilitate resolution of interface problems with other process owners • Ensure that the Capacity Management process, its associated techniques and methods are regularly reviewed and audited, and that all of these are subjected to continuous improvement and remain fit for purpose

Capacity Manager	This role has prime responsibility for ensuring that adequate IT capacity exists to meet required levels of service and for ensuring that IT management is correctly advised on how to cost effectively match capacity and demand. This role will oversee and coordinate implementation and maintenance of the Capacity Management process and activities throughout the IT organization. Key activities include:

- Produce capacity plans in line with business planning cycles; identifying requirements early enough to accommodate procurement and approval lead times
- Document need for increases and reductions in hardware based on service level requirements, targets and cost constraints
- Provide regular management reports which include current usage of resources, trends and forecasts
- Coordinate and oversee performance testing of new systems and solutions
- Provide holistic management of capacity that includes people and processes as well as technology
- Oversee decisions and actions to utilize Demand Management for controlling capacity when necessary
- Ensure shortfalls in capacity are recognized and appropriate corrective actions are identified and progressed on a timely basis
- Create and maintain forward looking capacity forecasts to predict future hardware/software spend
- Provide regular reports on capacity to the Availability Manager
- Organize and maintain a regular capacity review process
- Agree appropriate actions to maintain or improve capacity levels
- Initiate and coordinate actions required to maintain or improve capacity across IT units
- Act as a coordination point for changes to capacity levels when needed

Capacity Analyst	Primary role is to analyze capacity issues and problems to determine ways to improve capacity and performance at acceptable cost levels. Key activities include:

- Advise the Availability Management process about appropriate service levels or service level options based on capacity capabilities
- Size all proposed new systems to determine computer and network resources required taking into account hardware utilizations, performance service targets and cost implications
- Report on performance against service targets contained in Service Level Agreements
- Maintain a knowledge base of future demand for IT services and predict the effects of demand on performance targets and service levels
- Model impacts of changes in business volumes and new technologies to predict needed capacity
- Translate business events and drivers into IT workloads and volumes
- Determine performance targets and service levels that are achievable and cost justified
- Conduct ad-hoc performance and capacity studies on request from IT management
- Ensure requirements for reliability and availability are taken into account in all capacity planning and sizing activities
- Analyze and review actual Capacity levels achieved against SLAs and OLAs and UCs
- Provide a range of IT capacity reporting to ensure that agreed levels of capacity and performance are measured and monitored on an ongoing basis
- Determine the capacity requirements from the business for new or enhanced IT services
- Monitor actual capacity usage versus targets and to ensure shortfalls are addressed
- Participate in Change Control meetings to assess and authorize changes from a capacity perspective
- Conduct capacity risk assessment when needed
- Assist in SLA negotiation efforts from a capacity capability standpoint
- Define the key targets of capacity required for the IT infrastructure and its components that underpin a new or enhanced IT service as the basis for an SLA agreement
- Maintain an awareness of technology advancements and best practices that have impact on capacity
- Gather capacity data as needed for reporting and communications

Subject Matter Expert	Assists with analysis of capacity and performance issues, strategies and activities within area of expertise. Key activities include: • Assess new IT technologies for their relevance in terms of capacity, performance and cost • Work with development teams and operations personnel to recommend tuning of systems and make recommendations to IT management on the design and use of systems to help ensure optimum use of all hardware and operating system software resources • Interface between Capacity Management initiatives and IT support groups that underpin services • Provide expertise specific to a business or technology function • Provide expertise specific to hardware/software solutions and vendor specific solutions • Assist Capacity Manager in investigating assigned capacity issues through to resolution
Capacity Architect	Assists Capacity Management initiatives by providing overall strategy and design direction. Key activities include: • Recommend resolutions to performance related incidents and problems • Create capacity and performance design criteria to be applied to new or enhanced infrastructure design • Ensure IT services are designed to deliver the required levels of capacity and performance required by business units and customers • Ensure the levels of IT capacity required are cost justified • Identify capacity needs for the Business Units represented to IT • Document capacity related blueprints and designs as needed

IT Service Continuity Management Roles

Overview

The key roles involved with executing this process are summarized as follows:

- IT Service Continuity Management Process Owner
- IT Service Continuity Manager
- IT Service Continuity Team Leader
- IT Service Continuity Team Member
- Business Continuity Management Liaison

IT Service
Continuity
Management
Process Owner

Primary role is to supporting the overall Business Continuity Management process by ensuring that required IT technical and service facilities can be recovered within required and agreed business timescales. Key activities include:

- Define and deploy the IT Service Continuity Management process for execution and compliance across IT support organizations
- Co-ordinate inter-process changes with other process owners
- Provide process ownership through design, implementation and continuous improvement activities in the process life cycle
- Work with all functions within the IT community to ensure processes are executed as designed and measured accurately and completely
- Ensure alignment of the process to Business Continuity Management plans and activities
- Create a new environment by displaying a motivational impatience for results (be a behavior change leader)
- Publicize activities and build commitment and consensus around IT Service Continuity Management activities
- Facilitate resolution of interface problems with other process owners
- Ensure that the IT Service Continuity Management process, its associated techniques and methods are regularly reviewed and audited, and that all of these are subjected to continuous improvement and remain fit for purpose

| IT Service Continuity Manager | This role coordinates IT Service Continuity Management (ITSCM) activities between the existing Business Continuity Management organization and local Technical Centers to ensure timely recovery of IT services. Key activities include: |

- Ensure that ITSCM planning information is kept current and up to date used by all parts of the ITSCM process
- Maintain the ITSCM Plan document
- Develop and manage the ITSCM Plan to ensure that agreed recovery objectives of the business can be achieved
- Ensure that all local Technical Center IT Service areas are prepared and able to respond to an invocation of the Continuity Plan
- Maintain a comprehensive IT Continuity Testing Plan and Schedule
- Undertake quality reviews of all ITSCM procedures and ensure that these are incorporated into the Continuity Testing plans and schedules
- Participate in negotiation and management activities with providers of third party recovery services when necessary
- Support the delivery of IT services when invocation of continuity plans have occurred including:
- Coordination with crisis control teams
- Invocation of appropriate recovery facilities
- Resource management, direction and arbitration
- Recovery site management
- Return to normal site operations
- Participate in negotiations, agreement and maintenance of OLAs within the local Technical Centers that involve continuity processing services
- Participate in Change Control meetings to assess and authorize changes from an ITSCM perspective and ensuring proposed changes do not compromise ITSCM plans and capabilities
- Ensure appropriate OLAs in place to support continuity services
- Provide regular reports on continuity readiness, test results and other related issues to the IT Service Manager
- Initiate any actions required to maintain or improve IT service continuity
- Produce ITSCM plans in line with business planning cycles; identifying requirements early enough to accommodate procurement and approval lead times
- Develop and coordinate IT Service Continuity awareness, training and communication activities

IT Service Continuity Team Leader	Primary role is to represent a business unit or IT department by providing IT Service Continuity requirements and acting as a coordination point for conducting IT recovery actions in the event of a major business disruption. Key activities include: • Act as a single point of contact to coordinate requirements on behalf of the department or business unit being represented • Act as a single point of contact to coordinate recovery actions in the event of a major business disruption on behalf of the department or business unit represented • Maintain current skills and knowledge in recovery actions as dictated by the IT Service Continuity Plan • Provide input to the IT Service Continuity Plan on behalf of the organization represented • Participate in IT Service Continuity testing activities • Participate in IT Service Continuity awareness and communication activities when requested • Coordinate and integrate IT Service Continuity activities with Business Continuity Management activities at the business unit level
IT Service Continuity Team Member	Executes IT Service Continuity recovery actions in the event of a major business disruption as dictated by the IT Service Continuity Team Leader. Key activities include: • Ensure contact information is kept current and accurate • Execute recovery activities as dictated by the IT Service Continuity Team Leader in the event of a major business disruption • Assist in IT Service Continuity testing activities if requested
Business Continuity Management Liaison	Acts as a single point of contact for communicating Business Continuity Management (BCM) activities and plans. Key activities include: • Act as a single point of contact for BCM activities and plans • Provide feedback on compatibility of IT Service Continuity Management plans and activities with BCM plans and activities • Communicate inventory of Vital Business Functions to IT Service Continuity staff • Provide feedback on results of IT Service Continuity tests in terms of compliance with BCM goals and objectives • Communicate changes in BCM plans and activities to IT Service Continuity staff and management on a timely basis

IT Financial Management Roles

Overview	The key roles involved with executing this process are summarized as follows:

- IT Financial Management Process Owner
- IT Financial Manager
- IT Financial Analyst
- IT Financial Administrator
- Subject Matter Expert

IT Financial Management Process Owner	Primary role is to ensure executive support of the IT Financial Management process, co-ordinate the various functions and work activities at all levels of the process, provide the authority or ability to make changes in the process as required, and manage the process end-to-end so as to ensure optimal overall performance. Key activities include:

- Assist corporate management and finance organizations in developing policies for budgeting, accounting and charging
- Develop policies for monitoring and reporting on IT costs and charges
- Agree suitable IT accounting policies and procedures (i.e. how depreciation will be handled, what is cost versus capital, etc.)
- Communicate the organization vision and the process's strategic goals to business units and IT organizations
- Identify and assimilate similar and/or overlapping activities/ initiatives within IT
- Co-ordinate inter-process changes with other process owners
- Provide process ownership through design, implementation and continuous improvement activities in the process life cycle
- Work with all functions within the IT community to ensure processes are executed as designed and measured accurately and completely
- Ensure alignment of the process to the corporate and IT strategy
- Create a new environment by displaying a motivational impatience for results (be a behavior change leader)
- Publicize activities and build commitment and consensus around IT Financial Management activities
- Facilitate resolution of interface problems with other process owners
- Assist in communications and management of IT customer expectations of IT Services

| IT Financial Manager | This role is responsible for building and managing process that provide cost effective stewardship of IT assets and resources used to provide IT services. |

Key activities include:

- Develop IT account plans and investment cases when needed by organizations and business units
- Manage the IT organization budget
- Prepare IT budget forecasts and assist other organizations, when necessary, in preparing the IT elements of their budgets
- Report regularly on budget conformance to IT managers
- Identify budget conformance issues to IT managers
- Provide close support to Service Level, Availability, Capacity and IT Service Continuity management processes during budgeting and IT investment planning
- Produce IT financial plans in line with business planning cycles, identifying financial needs early enough to accommodate procurement and approval lead times
- Participate in Change Control meetings to assess and authorize changes from an IT Financial perspective
- Manage and operate IT billing and chargeback operations
- Resolve chargeback and billing disputes when they arise
- Conduct periodic internal audits of IT financial performance
- Assist with external audits of IT financial performance when requested by other corporate organizations
- Ensure IT charges and billings are accurate and truly reflect the services delivered

IT Financial Analyst	Support IT Financial Management activities with analysis of IT costs, budgets and chargeback operations. Key activities include:

- Advise Service Level Manager about IT costs and charges for service levels or service level options
- Identify suitable tools and processes for gathering IT cost data
- Develop suitable IT cost models
- Assist in developing cost/benefit cases for IT investments
- Advise management on the cost-effectiveness of IT solutions
- Assist with external audits when requested
- Provide support for IT service charging activities when necessary including:
- Identification of charging policies
- Providing justifications and comparisons for charges
- Preparing regular bills for customers
- Preparing price lists for services
- Conduct ad-hoc performance and IT Financial studies on request from IT management
- Analyze and break down IT infrastructures into cost components and categories
- Build and develop IT cost models
- Communicate the impacts of planned IT investments

IT Financial Administrator	Administrative support role to handle tasks that gather cost and budget data, administer financial reporting tools, maintain cost models, assemble and produce financial reports and any other tasks that assist the IT Financial Management process. Key activities include:

- Administer IT Chargeback and billing operations
- Gather cost and budget data when requested
- Assemble budget, accounting and charging reports
- Maintain Financial Management database

Service Desk Roles

Overview	The key roles involved with executing this function are summarized as follows:

- Service Desk Function Owner
- Service Desk Manager
- Service Desk Analyst
- Call Agent
- Service Desk Administrator
- Service Desk Infrastructure Architect

Service Desk Function Owner	Primary role is to ensure executive support of the Service Desk function, co-ordinate the work activities at all levels of the function, provide the authority or ability to make changes in the function as required, and manage the function end-to-end so as to ensure optimal overall performance. Key activities include:

- Communicate the organization vision and the function's strategic goals to business units and IT organizations
- Identify and assimilate similar and/or overlapping activities/ initiatives within IT
- Co-ordinate inter-function changes with ITSM process owners
- Provide function ownership through design, implementation and continuous improvement activities in the function life cycle
- Work with all functions within the IT community to ensure functions are executed as designed and measured accurately and completely
- Ensure alignment of the function to the corporate and IT strategy
- Create a new environment by displaying a motivational impatience for results (be a behavior change leader)
- Publicize activities and build commitment and consensus around Service Desk activities
- Facilitate resolution of interface problems with ITSM process owners
- Communicate with and manage the expectations of customers of IT Services

Service Desk Manager	Responsible for Management, Supervision, organization, customer satisfaction and staff support capabilities of the Service Desk. Key activities include:

- Provide leadership to Service Desk staff to develop and meet Service Desk goals and strategies
- Handle staff, customer and management concerns, problems and inquiries
- Provide call management and support services in line with Service Level targets
- Counsel and Coach Service Desk staff
- Perform periodic Service Desk staff performance reviews
- Review and analyze all Service Desk and Incident Management reports to proactively seek improvements
- Develop and communicate Service Desk staff training and skills maintenance plans
- Work with key Service Desk Customers to address customer related issues
- Oversee actions to obtain feedback from customers on IT service quality and communicate results to the ITSM Service Manager
- Administer and manage staffing levels in line with IT service needs
- Ensure all incidents are being addressed and appropriately escalated to support staff
- Manage all Service Desk staff recruiting activities
- Maintain Service Desk staff morale and strive for low staff turnover rates as much as possible
- Represent Service Desk functions on Change Management CAB meetings

Service Desk Analyst	Provide leadership and mentoring for Service Desk Call Agents to resolve Service Desk issues and maintain customer satisfaction at high levels. Key activities include:

- Mentor Service Desk Call Agents
- Recommend resolutions to technical problems reported by customers not easily resolved by Service Desk Call Agents
- Consult with peers on technical issues pertaining to all systems and applications
- Audit Incident Management Database to ensure all Incidents are being logged and categorized accurately
- Maintain high level of customer satisfaction
- Assist Service Desk Call Agents with appropriate support and escalation for reported incidents and requests within established service targets
- Act is a focal point for disseminating communications about services and policies among Service Desk staff
- Provide appropriate communications and turnover for long running incidents as needed
- Assist with actions to improve Service Desk services as requested
- Ensure all incidents are logged in the Incident Management Database
- Act as liaison between customer and user groups to support them in jointly resolving incidents
- Identify needs related to training, documentation, and technical issues
- Train new Service Desk staff members in Service Desk operating procedures
- Assist in preparation of staff schedules to ensure Service Desk is staffed appropriately to cover all hours of operation

Service Desk Administrator	Provides administrative support for Service Desk activities at the direction of the Service Desk Manager. Key activities include:

- Administer Call Management systems
- Collect and gather Call Management data for reporting purposes
- Assist with Service Desk incident communications as needed
- Maintain Service Desk escalation and contact lists
- Administer Service Desk Knowledge Bases as needed
- Administer changes to incident handling procedures as directed by the Service Desk Manager
- Prepare Service Desk training guides and materials with content provided by the Service Desk Manager
- Assist with other tasks as directed by the Service Desk Manager

Call Agent	Handles incidents and requests from customers and end-users to maintain high levels of satisfaction with IT services. Key activities include: • Maintain end to end responsibility for customer calls providing timely, reliable and courteous service • Provides customer service and first level technical resolution for operational and service-related incidents • Resolve or escalate incidents and requests in line with established Service Level targets • Support Security Management activities by exercising constant vigilance for possible security implications during customer interactions • Provide feedback of intelligence gained through customer interactions • Maintain appropriate level of skills to handle incidents and requests in line with established service levels • Log and record all reported incidents into the Incident Management Database • Respond to all customer requests with accurate and appropriate information • Identify improvements to Service Desk services and operation on an ongoing basis
Service Desk Infrastructure Architect	Designs and maintains the Service Desk infrastructure that is used to support Call Management and Monitoring activities. Key activities include: • Design and oversee development of Call Management infrastructure • Design and oversee development of monitoring infrastructure for the Service Desk • Recommend changes and improvements to Service Desk support tools • Liaison with 3rd party vendors and other IT organizations that underpin Service Desk functions • Resolve incidents related to Service Desk infrastructure errors

Other Roles

Overview

The key roles involved with executing this process are summarized as follows:

- ITSM Director
- Service Architect
- Technical Support Analyst
- Operations Manager
- Systems Administrator
- Operations Analyst
- Scheduler
- Storage Manager
- Database Administrator
- Asset Administrator
- License Administrator

ITSM Director

This role oversees the entire IT Service Management (ITSM) operation to ensure that quality service management processes are developed and deployed to meet agreed business objectives. Key activities include:

- Manage ITSM staff and budget
- Enable and champion an IT service culture
- Develop, implement and maintain ITSM-based management processes and controls to ensure quality is maintained to meet business objectives
- Champion and promote service improvements on an ongoing basis to continually improve quality and customer satisfaction with IT services
- Maintain day to day responsibility for the ownership and resolution (including any referral or escalation as may be necessary) of Service Management issues which arise in connection with ITSM Services
- Review service metrics (KPIs) that identify the success of the services being utilized to recommend and coordinate implementation of changes to ITSM services to improve metrics
- Work to ensure continuous alignment of the services with customer needs, i.e. changing work patterns, workloads, revised objectives
- Co-ordinate inter-process changes with ITSM process owners
- Ensure alignment of ITSM solutions to the corporate and IT strategy
- Create a new environment by displaying a motivational impatience for results (be a behavior change leader)
- Publicize activities and build commitment and consensus around ITSM

Service Architect	This role is responsible for the overall coordination and design of the ICT infrastructure. Key activities include:

- Design a secure and resilient service infrastructure that underpins ITSM solutions and activities
- Maintain all service design, architectural, policy and specification documentation
- Identify opportunities for automation for manual and redundant operational tasks
- Recommend service solutions to continually improve ITSM solutions on an ongoing basis
- Identify service design policies, philosophies and criteria covering connectivity, capacity, interfaces, security, resilience, recovery, access and remote access, ensuring that all new services adequately underpin ITSM solutions and activities
- Provides advice and guidance to analysts, planners, designers and developers on all aspects of service design and technology
- Interface with designers and planners from external suppliers and Service Providers, ensuring all external service services are designed to meet their agreed service levels and targets
- Review and contribute to the design, development and production of new services, SLAs, OLAs and UCs, covering service equipment and services
- Maintain a good technical knowledge of all installed service product capabilities and the technical frameworks in which they operate

Technical Support Analyst	Support service infrastructure solutions with technical skills expertise to maintain steady state service operations. Key activities include:

- Analyze service infrastructure technical issues
- Resolve service infrastructure incidents and problems when they occur
- Maintain Ownership of problem diagnosis, resolution and escalation for all received problems and issues
- Participate in Design and Planning for new operational services when needed
- Apply third party maintenance to service infrastructure components
- Develop and maintain service infrastructure documentation and procedures
- Provide technical support and assistance with feasibility studies

Operations Manager	This role manages day-to-day service infrastructure operations to provide delivery of systems and services in order to meet or exceed agreed services levels. Key activities include:

- Single point of ownership for effective provision of systems and services to customers
- Oversee operational activities and services for one or more operational delivery centers
- Proactively identify and implement service improvements in operational delivery centers
- Approve acceptance into production of new systems and services
- Maintain the Operations Technical Library
- Ensure BCP/DR plans are compatible with operational delivery center operations and that plans are tested on a regular basis
- Represent interests of operational delivery centers managed for all ITSM service initiatives
- Manage and develop all service operational delivery center staff
- Oversee recruitment of operational staff
- Manage operational delivery center budget
- Approve procurement of new CIs that will reside in the operational delivery center
- Provide service reports on operational delivery center performance and quality

Systems Administrator	This role administers and maintains infrastructure devices such as servers, hosts and networking devices to ensure proper operation and availability. Key activities include:

- Control and administer hardware and operating software configurations
- Monitor devices for proper operation and performance
- Apply vendor provided maintenance to devices
- Detect, diagnose, isolate and correct device operational failures
- Support Incident and Problem Management activities with device expertise and troubleshooting
- Maintain awareness of new technologies that might enhance device operation, capacity and performance

Operations Analyst	This role is responsible for performing all operational processes and procedures, ensuring that all services and infrastructure meet their operational targets. Key activities include:

- Operate and implement all operational service infrastructure and procedures
- Participate in incident and problem support activities when requested
- Investigate, diagnose, and take prescribed actions on all operational events, alarms and incidents
- Monitor all service Operations and services to ensure service quality is being delivered on a daily basis
- Maintain operational logs and journals on all events, warnings, alerts and alarms, recording and classifying all messages; maintain all operational data collection procedures, mechanisms and tools
- Maintain all operational documentation, processes, management and diagnostic tools and spares, ensuring that spares are maintained at the agreed levels
- Ensure that all routine maintenance tasks are completed on all operational infrastructures
- Ensure that all infrastructure equipment is maintained according to policies and recommendations and perform regular checks on environmental equipment and conditions

Scheduler	This role is responsible for the management and control of all aspects of the scheduling, monitoring and control of operational workloads. Key activities include:

- Prepare and maintain day-to-day service workload schedules in line with scheduling guidelines
- Ensure that operational workloads are run according to their defined schedules
- Process ad hoc workload requests when requested and approved
- Administer scheduling tools and infrastructure
- Develop and maintain all necessary operational scheduling documentation
- Produce workload scheduling reports that report results of schedules and job runs in a timely fashion

Storage Analyst	This role is responsible for the management and control of storage media, backup and recovery schedules, testing, storage planning, allocation, monitoring and decommissioning. Key activities include:

- Interface with Availability, Capacity, Security and IT Service Continuity Management to ensure that all requirements are met by current backup and recovery policies
- Develop and manage a Data Retention Policy that is compliant with legal and regulatory requirements
- Implement and administer backup and recovery packages and tools
- Procure magnetic tapes, diskettes, cartridges, paper, microfiche and all other media and devices when required
- Manage and maintain media pick lists and vaulting mechanisms
- Establish and maintain a clear physical identification system for media for easy identification
- Monitor backup jobs and schedules to ensure these take place without error

Database Administrator	This role is responsible for the management and control of databases in the infrastructure. Key activities include:

- Perform database physical and logical design tasks to meet the objectives new IT solutions and services
- Implement database logging and recovery operations
- Work with Capacity Management to perform database sizing and support capacity workload and forecasting estimates for databases
- Implement database backup and recovery procedures
- Monitor databases for adequate performance and capacity
- Assist IT developers with database architecture and access control policies
- Identify appropriate database solutions and products

Asset Administrator	This role is responsible for tracking and control of physical IT assets in the infrastructure. Key activities include: • Handle procurement of IT assets when requested • Tag and track all IT assets, their locations and owners • Administer the Asset Management Database and the Asset Inventory • Perform asset disposal tasks in line with corporate asset policies • Perform periodic asset discovery and audit tasks • Receive assets and ensure delivery to correct locations • Coordinate asset setup and teardown activities when requested
License Administrator	This role is responsible for tracking and control of software licenses in the infrastructure. Key activities include: • Handle procurement of software licenses when requested • Track all software licenses, their usage and owners • Administer the License Management Database • Perform procurement tasks for software licenses • Assist with software license audit activities when requested • Ensure licenses are compliant with vendor usage specifications • Interface with Release Management activities and policies

ITSM Design Principles

*"It is the nature of man as he grows older, a small bridge in time, to
protest against change, particularly change for the better."*
—John Steinbeck

This chapter briefly describes the concepts of using Guiding Principles when designing ITSM solutions. It then presents some examples of Guiding Principles that can be used by Implementation teams when building ITSM solutions.

Guiding Principles are statements about how IT Services should operate. They support the IT Service Management Vision. Guiding Principles are critical statements of direction that will have major impact on how IT services should be designed and operated. Guiding Principles should be clearly understood and communicated both internally and externally to IT Services. They are derived from a combination of:

- Basic beliefs
- Experience
- Company priorities
- Underlying culture within a business organization
- People involved with the delivery of IT Services.

Guiding Principles have three parts to them:

- A Statement
- A Rationale Description
- A List of Implications

The Statement consists of a one sentence statement that states the principle. For example:

We will let our customers know the key services we offer them and who is accountable.

The Rationale Description lists reasons for why the principle should be accepted by the business organization. Things to consider when documenting the rationale include:

- Why should the organization do this?

- What business benefits does this Guiding Principle advance?

- What characteristics can be used to defend the Guiding Principle?

The Implication List describes areas of impact to business and IT units as a result of operating with the principle. Things to consider when documenting implications include:

- What needs to be done if the Guiding Principle is implemented?

- What impact will it have on business and IT units?

- What kind of behaviours, tools, data or processes need to be in place?

Well-designed Principles have the following characteristics:

- They state a fundamental belief of the enterprise in one or two clearly written sentences.

- They have relevance to the IT Service Management processes.

- They are worded directly and simply in terms understandable by both business and IT managers.

- They have wide applicability.

- They are durable and will not be outdated quickly by advancing technology.

- They have impacts which need to be documented.

The following are examples of some well-defined Principles:

- Customer information will be kept strictly confidential within policies set by the organization and regulatory agencies.

- Service Management solutions, whether purchased or developed internally, will be highly structured and modular.

Characteristics that make up poorly designed Principles are:

- Statements that are difficult to dispute

- Is a general business or financial statement.

- Does not support business goals

- Is stated at too low a level or names a product/technology.

- May be included because management says so

The following are examples of some poorly defined Principles:

- The overall cost of computing needs to be reduced.

- All servers will use the EISA Bus to achieve high performance.

- Only Ethernet LANs will be implemented in our corporation.

The following pages present examples of Guiding Principles. The purpose of these examples is not to dictate recommended Principles, but to give a starting point and stimulate some thinking for what Principles might be used unique to the business.

Overriding Principle Examples

Guiding Principle	Rationale	Implications
We will consider the benefit and impact on the customer of everything that we do.	• Helps IT Services to understand customer needs. • A customer-focused services organization ensures business aligned IT services. • Enables a customer-focused culture. • Increases customer satisfaction.	• IT Services understands the implications of actions they take on the customer and business. • IT Services understands the business calendar. • Understands legislation and other influencing standards. • Activities are measured in terms that are relevant to the business.
Insource or outsource decisions will be based on a clearly defined set of criteria.	• Ensures IT services personnel focus on activities with the greatest value to the business. • Provides IT services personnel with a greater sense of responsibility for the business' success. • Emphasizes value activities are in the business, not IT. • Enables IT suppliers to deliver more cost effective solutions or add greater value in areas of their core competencies.	• Analysis is done to determine the business critical areas for IT services. • Liaison with third parties is done to ensure non-critical services are delivered seamlessly. • Determined critical and non-critical services. • Established procurement process for services. • Process is in place to manage Service Levels and ensure smooth fit with internal services. • Improvement case guidelines for standard preparation of cost or service are in place. • Decision making criteria is defined and maintained.

Guiding Principle	Rationale	Implications
All IT services activities will be considered in the context of the business strategy.	• While we are focused on our customers' needs we need to reconcile these with the business imperatives. • We are a cost to the business; we shouldn't do anything that does not directly support the business. • We need to clearly demonstrate our value to the business. • We must position ourselves to add maximum value where business has a clear strategy with IT consequences, such as convergence on ERP.	• Service requirements are anticipated in time to meet sudden business moves. • The business and IT strategy has been clearly communicated and understood by the IT services personnel. • The IT strategy is updated as required. • A process to ensure continued linkage of strategies is in place. • The need to arbitrate between conflicting customer and business demands has been considered.
We will regularly measure and report on the services we provide.	• If we can't measure it, we can't manage it. • Demonstrate the value provided by the services we offer so the customer will understand the investment needed to improve the service. • Enables timely, quality service provision.	• A process is in place to capture the pertinent data and report it a regular basis. • Measures are agreed to with the customer. • We understand that measuring is not enough and we have the disciplines and processes in place to act on issues raised. • We understand that measurement simply provides data and we have established methods to turn the data into knowledge (e.g. trend and threshold analysis). • The proactive cycle of measure, analyse, report and action is part of our management culture.

Guiding Principle	Rationale	Implications
IT Service Management activities will be proactive rather than reactive, whenever possible.	• Promotes highest possible service levels. • Allows IT reaction before users are involved in many cases. • Contributes toward increased customer satisfaction.	• We find tools for all processes. • Requirements for a service orientation have been met. • We reward people who prevent incidents more than fixing incidents. • We take proactive approaches towards preventing incidents.
Our people are our primary asset—we will motivate, develop and retain them.	• Staff retention is much more cost-effective than staff recruitment. • Business awareness is grown, not taught. • No amount of documentation can replace experience. • Motivated staff tends to be loyal staff. • Personal development is key to motivation.	• Objective HR measures are in place. • We understand what motivates our people. • Staff development costs are understood. • We increase performance management; we do not accept mediocrity.
Each systems management process will have a single process owner, responsible for process quality and integrity.	• Keeps the process discrete and bounded • Avoids responsibility conflict or uncertainty. • Promotes responsibility and continuous improvement.	• We ensure the process owner is at the right level in the organization. • We determine availability of and invest in dedicated owners. • The process owner helps define the process ownership role. • We have appropriate sponsorship within IT or the business. • Some process owners may come from operational support groups.

Guiding Principle	Rationale	Implications
IT services will focus on proactive, end-to-end, automated management through an integrated Enterprise System Management Framework.	• Accepted research and industry experience shows that the deployment of an ESM framework delivers maximum gains as a scalable service environment emerges. • A consistent approach and interface for all aspects of services delivery allows greater flexibility in deploying support personnel and accommodating new areas of service. • Ensures that tool expenditures are focused within an overall management solution that avoids redundant tool functionality. • Reduces costs for management and maintenance.	• Management of internal OLA's and external SLA's is in place to provide end-to-end service management. • Audit of current tool-set is done to ensure that deployment of the framework is conflict free. • Transformation plans are in place to ensure smooth migration to the framework. • Review of current solutions and environments is done. • There are "willing to live with" solutions that may not be best of breed individually, but provide greater overall value through integration.
We will discover and incorporate industry best practice and tools.	• Our business organization is not a developer of IT service tools and methods. • Technical capability of products is developing very quickly. • Why re-invent the wheel? • Borrow and adapt best practice to meet our needs. • When we work smart the business benefits.	• Our culture is ready to embrace change. • Sound external advice is seen as an investment. • We recognize that research and innovation are part of our job.

Guiding Principle	Rationale	Implications
Standard systems management products will be used wherever possible, with minimal modifications and in-house code.	<ul><li>Reduces maintenance costs.</li><li>Allows staff to concentrate on implementing new functions.</li><li>Allows easier migration to updated tools.</li><li>Coping with new technology impacts will not be responsibility of IT.</li><li>Custom solutions usually involve higher costs in the long term.</li></ul>	<ul><li>We concentrate effort to analyze any demands for modifications to ensure they are required.</li><li>We sometimes sacrifice specific tailored function for long-term gain.</li><li>We look to use standard products when available.</li></ul>
IT Service Management processes must use common data wherever possible.	<ul><li>Promotes integration and correlation of data.</li><li>Reduces the need for re-entry and chance for error.</li><li>Simpler to administer the data.</li></ul>	<ul><li>We utilize powerful automated management systems.</li><li>We strive to obtain availability of integrated data solutions whenever possible.</li><li>Data design and normalization activities are done to accomplish this goal.</li><li>Data is interfaced or ported between different automation solutions.</li></ul>

Process Principle Examples

Incident Management		
Guiding Principle	**Rationale**	**Implications**
Incident processing and handling must be aligned with overall service levels and objectives.	• Ensures incident handling supports service levels and objectives. • Identifies which incidents require high priority handling based on actual business need.	• We understand service levels and objectives that need to be provided for. • We build event handling that is targeted towards service levels. • We develop Service Desk and Incident Management Operational Level Agreements to set thresholds for when Incident escalation and notification actions should be taken. • We implement reporting on service actions and thresholds within Incident Management functions. • We time stamp key Incident Management activities.

Problem Management		
Guiding Principle	**Rationale**	**Implications**
Every high severity incident will require a documented Root Cause.	• Ensures Problem Management focus on all high severity incidents. • Proactively works to reduce or eliminate future high severity incidents. • Provides upper management communication and assurances that high severity incidents are being appropriately addressed.	• Problem Owners are assigned to every SEV1 generated • Solid skills are in place to identify Root Cause • Management communications are in place to communicate timely and useful Incident information

Guiding Principle	Rationale	Implications
Problems will be tracked separately from Incidents.	• Provides clear separation between proactive Problem Management activities and reactive Incident activities • Easier ability to track Problem management activities and progress separately	• Tools are in place to track Problems separately from Incidents • Problem categories and reporting are in place
Every problem will have an assigned owner.	• Ensures responsibility assigned to fix problems • Provides single point of contact for communications about problems	• The Problem Owner role is understood and communicated throughout the organization • Problem Owners have appropriate authorization to coordinate and take actions to identify Root Cause and remove the error

Change Management

Guiding Principle	Rationale	Implications
No changes will be made to production systems without the approval of the IT and user representatives responsible for the service.	• Promotes an emphasis on service availability. • Ensures changes are only made with user involvement. • Ensures changes are authorized.	• We define user representatives and service owners. • We agree on change criteria. • We utilize an efficient process for approvals. • We have strong management support for the change process.

Release Management

Guiding Principle	Rationale	Implications
Operations will require sign-off on the acceptance criteria for new applications at the business study, design and pre-implementation stages with service delivery veto.	• Ensures that the system will meet business unit quality expectations. • Ensures application integrity. • Reduces production operational issues at the outset of production.	• We understand the potential for delayed implementation. • Resource skill levels needed to review solutions are understood.
Changes to the client environment will be minimized by adopting a Release Package concept.	• Reduces outage potential. • Minimizes costs of implementing releases. • Promotes release planning.	• Investment in a test facility for a group of releases. • Requires buy-in by release implementers. • Ability to plan releases.

Configuration Management		
Guiding Principle	Rationale	Implications
The configuration database will contain administrative data, and no operational data, which will be will be separate from the administrative databases.	• Efficient Configuration Management Database (CMDB), most notably performance. • Protects operational data.	• We understand that underlying databases will always employ different technologies.
The Configuration Management Database (CMDB) will consist of pointers and linkages to other company repositories versus having everything stored in the database itself.	• Allows owners to manage other repositories with minimal changes. • Avoids complex and overly large CMDB architecture.	• We determine how interfaces to repositories will be implemented. • We extend Configuration Management practices to groups that own the repositories. • We make sure repositories are adequately managed and kept current.
The CMDB will contain all infrastructure information.	• Centralizes all needed infrastructure where it can be easily found. • Provides recognition for authorized sources of infrastructure information. • Easily managed and controlled.	• Locate and find all authorized sources of infrastructure information. • Have mechanisms within the CMDB to link to physical items such as documentation manuals. • Have a CMDB database schema that accommodates a wide range of items such as organizational policies and design documents.

Service Level Management

Guiding Principle	Rationale	Implications
We will let our customers know the key services we offer them and who is accountable.	• Clear accountability for customer service. • Clarity for the customer.	• A service catalog that defines each service is in place. • Cross-organizational capabilities are in place. • Approaches and responsibilities are publicized. • Each service has a single owner. • Roles and responsibilities are strongly communicated. • A process is in place to maintain information on the accountability structure and ensure this information is current for customers. • An account management role has been established and clearly communicated. • Conflicts between owners and account managers are managed.

Service Level Management

Guiding Principle	Rationale	Implications
We will measure and deliver services based on Key Performance Indicators (KPIs) agreed to with the business.	• Enables the business to understand the value of IT services. • IT services can better communicate the business impact and benefits from services offered.	• IT services understands the business use of the services. • Service Level Agreements are in place to ensure both parties have equal expectations of the service to be delivered. • A mechanism for measurement and regular reporting on these services is in place. • KPI definitions that utilize business terms are in place. • Communication of KPI's and business value is done. • A process to track KPI's and adjust the service accordingly is in place.

Guiding Principle	Rationale	Implications
Service reporting to the client will be business based, which will include sub-reporting of services comprising the environment.	Provides a business-oriented picture of real achievement by the IT supplier.	• High levels of implementation work for reporting systems. • Feeds into and links to the charging system. • Data and metric availability.

Service Level Management

Guiding Principle	Rationale	Implications
We will take action to continuously improve service quality.	• Continuous improvement is an added value. • Simply measuring and reporting does not ensure continuous improvement in our services. • Objective measurement facilitates cost-benefit analysis of changes.	• We have objective measurements. • We educate everyone involved in the improvement cycle. • Our management processes are focused on change. • We understand that improvement activities must be seen as an investment, they in cost time and resources, which are recovered later in terms of quality of service delivery. • We communicate to all the benefits of continuous improvement. • Our culture recognizes that continuous improvement of service quality is a key part of management's role.

Availability Management

Guiding Principle	Rationale	Implications
Services will be managed on an end-to-end basis.	• Accountability • Improved orientation to the business • Improved quality of services delivery	• Changed organizational structure • Changes in service managers' responsibility and accountability (increased commercial awareness) • Increased education and training requirements • Ability of existing technology support this principle (legacy constraints) • Impacts on technical architecture • Greater control to the end users • Issues for application design & development
We will be the guardians of IT systems and services, ensuring they are available in-line with business needs.	• IT Supported systems are key enablers. • System availability is the way that IT adds value to the business. • Without IT systems the business data is inaccessible. • There is a direct cost to business for unplanned downtime.	• The business is enshrined in our SLA's. • We are prepared to enable rather than inhibit business change. • We fully understand the business requirements. • Our organization is focused on availability.
We will be the custodians of all company data, ensuring it is correctly stored, readily available, secure and recoverable.	• Customers feel secure that their business data is secure and stored in backups, ready for recovery. • Data is a critical asset with high monetary and business value.	• We have clear definitions of company and personal data. • Supported platforms are published with guidelines for where data is stored. • We are clear on integrity. • We have implemented security guidelines. • Data management guidelines have been put into place.

Availability Management

Guiding Principle	Rationale	Implications
We only will manage and support IT assets procured through IT.	• Ensures clear support responsibilities for non-standard or personal IT equipment. • IT services delivers better service by focusing on standard equipment where IT knows about the equipment operating environment and characteristics.	• Clear and unambiguous guidelines have been provided regarding the procurement of IT assets. • Review of existing and legacy systems and solutions is in place. • IT Enterprise architecture forms the basis for procurement decisions and standards. • Correct procurement channels have been defined and put into place.
Outside hours of committed system or service availability, clients will receive no guarantee of availability of such services.	• Minimize costs of support. • Ensure support is available when needed. • Ensure clear expectations of support availability requirements.	• Effective interfaces to service level management needed to ensure needs are defined. • Exception criteria defined to allow for special needs.

Capacity Management

Guiding Principle	Rationale	Implications
Business units will provide consolidated business capacity planning information.	• Promotes business unit buy-in and responsibility for influencing the capacity requirements for their services. • Promotes a business-oriented approach to defining capacity planning information. • Improves the reliability of the planning data. • All capacity usage is driven by business events.	• Business units have committed to the process. • We have a method of translating business data into IT terms.

IT Service Continuity Management

Guiding Principle	Rationale	Implications
There will be an annual test of IT Service Continuity recovery processes.	• Ensures that critical data and systems can be quickly recovered in the event of a major outage. • Ensures that services recovered match those currently in production. • Ensures that capacities for recovery are truly effective.	• Management is committed to the yearly test. • We understand testing expenses. • We perform ongoing analysis of what is critical to the business.

Financial Management

Guiding Principle	Rationale	Implications
All service offerings will be reviewed at regular intervals to determine whether they are economically justified from both the provider's and customer's points of view.	• Assures the affordability of services offered/received. • Identifies services with low value and high support costs.	• We have tools that allocate costs appropriately. • Only valuable services are funded. • We understand what services IT is offering. • We understand IT delivery costs and the service delivery chain.
Users will be charged for IT services in a manner that reflects the true costs of using them.	• Focuses on the true cost of IT services. • Controls user demands. • Promotes justification mentality.	• We understand that definition of charging mechanisms is a challenge. • We have support of the enterprise. • We have clear charging policies (i.e. profit center, cost recovery only, etc.).

Other Principle Examples

Service Desk		
Guiding Principle	**Rationale**	**Implications**
All company employees will have toll free access to the company Service Desk.	• Ease of access to services. • Promotes use of the service interface. • Reduces the number of interfaces. • Promotes user satisfaction. • Provides single-point-of-contact for all IT services.	• Service Desk support is effective. • There is investment in tools to provide support. • User satisfaction is continually measured. • Telecom costs have been considered. • Overseas calls are accommodated.

Security Management		
Guiding Principle	**Rationale**	**Implications**
IT is responsible for the integrity of data associated with IT services, irrespective of technical platform.	• Data is a corporate asset which IT is mandated to manage. • Users will not do it. • There needs to be separation between producers of data and those managing it.	• We make users aware of what data management IT will and will not provide through SLA's. • We are able to handle increasingly complex data management requirements driven through increased business demands and regulations.

Security Management		
Guiding Principle	**Rationale**	**Implications**
No sensitive data will reside on client workstations.	• Minimizes exposures of data being compromised or lost. • Reduces security, backup and recovery complexity and costs.	• We perform regular workstation audits. • We have simplified backup procedures and given proper data staging rules. • We have identified where to stage and locate sensitive data.

Operational Principles

Guiding Principle	Rationale	Implications
There will be frequent tests of backups.	• Data and systems must be proven to be protected. • Needs change and thus the protection mechanisms must evolve and be tested as workable. • The business relies on the IT services and data contained in IT devices.	• Defined recovery procedures and defined critical data. • Management commitment to the costs of testing. • Updated recovery processes.
Backup and recovery processes will be completely automated.	• Ensures ability to recover key data. • Ensures high service levels. • Avoids client involvement in the task (and the possibility of inattention to the task).	• Cost of implementation and testing. • Affects availability schedule. • Application dependencies.
Event notifications will only go to those responsible for handling or decision processes related to them.	• Directs event handling activities only to those who must process them. • Avoid needless notifications to those not directly involved in processing events.	• We identify which departments, groups or individuals need to respond to events. • We maintain event routing information as new events are added or personnel responsibilities change.

Operational Principles

Guiding Principle	Rationale	Implications
All event management and support will be centralized.	• Avoids conflicts in management of events. • Allows for appropriate operational response for new events and changes in processes. • Avoids support personnel receiving event notifications for events they are not prepared to handle.	• A common rule base has been built and maintained. • We process rule changes and additions through a change management process. • Support personnel have buy-in processes to accept/reject event changes.

Operational Principles		
Guiding Principle	**Rationale**	**Implications**
All application events must utilize a common set of messaging and logging standards.	• Consistent, common ways to process and handle events. • Faster implementation of event processing. • Sets common expectations for how events will be recognized and handled.	• Publish a standard set of APIs, event message formats, usage and event classification criteria. • Communicate event management standards to application development personnel.
Only application events that are made available will be recognized and processed.	• Cannot effectively monitor what cannot be seen.	• Application developers identify which application events are to be recognized. • Application developers build messaging hooks into the applications for events they wish to recognize.

Operational Principles		
Guiding Principle	**Rationale**	**Implications**
Event processes will be automated whenever possible.	• Eliminates potential problems that can be caused by human error. • Targets event processing that can occur transparently to end-users and IT personnel.	• We implement programs to continuously look for candidates for automation. • Automation solutions for application events involve developers who may need to construct additional scripts, code or procedures to provide automated recovery at the applications level. • Application dependencies are considered when developing automated response handling.

Chapter

16

Continual Service Improvement

"Continuous improvement is better than delayed perfection"
—Mark Twain

ITSM transformation does not stop once ITSM solutions have been put into place. It takes an ongoing effort to continually monitor, manage, improve and update IT services to stay current with business needs and evolving technologies. Establishing a Continual Service Improvement (CSI) program is an ongoing activity that will formalize how service improvement recommendations will be recorded, registered and acted upon. This means that a standard means needs to be in place for recording enhancement requirements, selecting ones to fund and work on, tracking and approving IT Service improvement requests and projects and then linking these projects to the IT service portfolio.

CSI Program Overview

The key objective of the CSI activity is to establish a standard means for recording enhancement requirements, selecting ones to fund and work on, tracking and approving IT Service improvement requests and projects and then linking these projects to the IT service portfolio. The goal is to make IT services better and better over time as well as keeping them relevant with current customer needs. Key activities to accomplish this objective are:

- Establish metrics, reporting, analytical methods and resources to continually review service quality and delivery issues

- Establish an ITSM Continual Service Improvement (CSI) Register to record and track improvement requests and projects

- Establish a means and a process for prioritizing, selecting and approving ITSM improvement projects

- Implement tracking and reporting for ITSM improvement initiatives and their progress

Overall, this effort allows a means for service management executives to:

- Identify and raise IT service improvement initiatives

- Utilize a fair and equitable approach for prioritizing and selecting the appropriate initiatives to move forward with

- Obtain transparency in what initiatives have been approved, their status and progress

Similar to the ITSM transition stages discussed earlier, CSI is initially established with a project effort that combines people, process and technology to put into place. A high level view of activities to put CSI into place looks as follows:

Technology Efforts

- Establish a CSI registry to record and report on IT service improvement initiatives

- Establish a weighting scheme to score and prioritize initiatives based on their benefits, risk and effort to implement

Process Efforts

- Establish a CSI process and set of activities

People Activities

- Establish CSI governance roles and responsibilities

- Establish resources to review data analyzing for activities that may be undertaken to improve services

Data

- Identify relevant data for analyzing the quality of the service provided along with any related service delivery issues

- Identify relevant data for tracking and reporting on CSI initiative status and progress

- Establish CSI reports and progress metrics

CSI Project Implementation

- Establish an ITSM CSI Register and process

- Define the Continual Service Improvement Process

- Define CSI Roles and Responsibilities

- Populate the CSI register with Inventory of CSI requests / initiatives

- Establish a process for selecting and approving ITSM improvement projects

- Establish CSI Project Approval Gates and criteria

- Establish CSI project Portfolio Modeling Capabilities

- Establish Criteria For Prioritizing Investments In CSI Projects

- Implement tracking and reporting for ITSM improvements

- Establish a portfolio of CSI projects

- Build CSI Reports

- Build Project Portfolio Reports

The CSI project efforts can be considered complete when the following has been established:

- A Continual Service Improvement Process has been established and put into place

- CSI Roles and Responsibilities have been defined with resources allocated to them

- A CSI register is in place and populated with an initial inventory of CSI requests and initiatives

- An initial portfolio of CSI projects has been established

- CSI reports and metrics have been put into place

- Project Portfolio Reports have been put into place

- Regular meetings and decision making have been put into place

- Communications have been issued about the new CSI program and how it operates.

Project Team

The table below lists recommended roles for operating a Continual Service Improvement program. The CSI function itself may exist in a separate organization or as a virtual organization that draws from different IT departments and units. The CSI Resource teams may also be part of this organization, but usually will reside within IT line functions.

Project Team Role	Key Project Activities
ITSM Steering Committee	Oversees Service management efforts, funds and prioritizes CSI projects and initiatives
CSI Project Director	ITSM Program Manager accountable to Project Sponsor & ITSM Steering Committee
CSI Project Manager	Responsible for day to day project management & guidance to Project Lead
CSI Project Coordinator	Assists Project Director, Project Manager
CSI Architect	Builds the CSI Register and reports; populates the CSI register
CSI Process Lead	Designs, builds and publishes CSI control processes, procedures and work instructions
CSI Resource Teams	Responsible for hands-on activities to identify opportunities for service improvement and to make those service improvements once authorized

CSI Process Overview

The Continual Service Improvement process supports execution of Deming's Plan-Do-Check-Act Cycle for a quality service improvement cycle. The recommended process is cyclical in that it repeats itself on a periodic basis (typically 3-12 months).

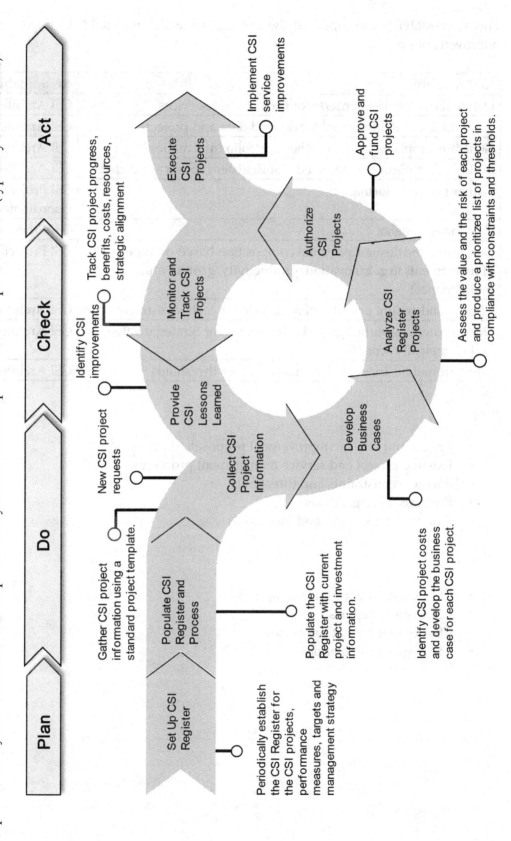

Set Up CSI Register

This step establishes (or updates) the CSI register which will hold the portfolio of service improvement projects.

Key Activity	Accountable Role
Define the information needs for projects and services	CSI Architect
Customize CSI Register tool for desired data and projects	CSI Architect
Generate templates used to gather portfolio information	CSI Architect
Build and configure customized CSI Register reports as needed	CSI Architect
Gather project information	CSI Project Coordinator
Categorize benefits	CSI Process Lead
Determine the desired portfolio composition based on types of investments (e.g. Innovation, productivity, growth and maintenance)	CSI Project Director
Choose and weight prioritization criteria for each investment type	CSI Architect
Define criteria, weightings and calculations for Strategic Value and Complexity scores	CSI Architect
Input annual business constraints and other thresholds	CSI Architect

Inputs

- Project and service improvement proposals
- Existing project and service management processes
- Business constraints and thresholds
- Business Strategic Plans
- Financial constraints and thresholds

Outputs

- CSI Register information templates
- CSI Register reports
- CSI Register benefit categories
- CSI Register calculations

Populate CSI Register and Process

This step populates the CSI Register with projects and project investment information. It also develops (or updates) the CSI process itself.

Key Activity	Accountable Role
Develop, review, approve and maintain decision criteria (Decision Criteria consist of: Qualification or Screening, and Evaluation or Prioritization	CSI Architect
Develop and maintain the CSI process	CSI Process Lead
Populate or update CSI Register with current service improvements and projects	CSI Project Coordinator
Request updates to strategic and tactical plans, as required	CSI Project Director

Inputs

- Lessons Learned
- Existing CSI Register or Service Improvement Projects
- CSI Investment Status
- Project and Service Performance
- Service Strategy and Directions
- Market, Industry and Technical Trends

Outputs

- CSI Decision Criteria
- CSI Qualification or Screenings
- CSI Evaluation and Prioritizations
- CSI Process Flow and Documentation

Collect CSI Project Information

This step identifies service improvement opportunities and places those into the CSI Register.

Key Activity	Accountable Role
Receive requests for service improvement efforts	CSI Project Coordinator CSI Resource Teams
Gather current improvement project and service status	CSI Project Coordinator
Review service progress and attainment	CSI Project Coordinator CSI Resource Teams
Identify key improvement project and service issues	CSI Architect CSI Resource Teams
Identify project/service mitigation actions	CSI Architect CSI Resource Teams
Translate actions into additional service improvement projects	CSI Architect CSI Resource Teams
Update CSI Register with candidate projects	CSI Project Coordinator
Provide CSI status and performance reports to management	CSI Project Coordinator

Inputs

- IT Service Capabilities
- Service Improvement Project Histories
- Customer and Market Intelligence
- Industry and Government Standards and Regulations
- CSI Project Status Reports
- CSI Project Work Plans and Schedules
- CSI Project Financial Reports
- CSI Issues/Action Item Logs
- Service Costs and Issues

Outputs

- CSI Project Performance Summaries
- CSI Project And Service Costs—Planned Versus Actuals
- Summary Of CSI Project Or Service Issues
- CSI Register Updates

Develop Business Cases

This step develops the business case for each CSI project.

Key Activity	Accountable Role
Identify resources and requirements for CSI projects and project changes	CSI Project Manager CSI Resource Teams
Identify CSI project and service business cases and cost estimates	CSI Project Manager CSI Resource Teams
Develop CSI project and cost proposals	CSI Project Manager
Develop Business Case Document	CSI Project Manager
Develop Business Case Presentation	CSI Project Manager CSI Resource Teams
Develop Cost/Benefit ROI Model	CSI Project Manager
Evaluate Investment Proposal against Investment Screening Criteria	CSI Project Director
Decide if Investment Proposal qualifies for possible inclusion in the CSI Register	CSI Project Director
Review Business Case recommendation	CSI Project Director
Validate Business Case with current environment	CSI Project Director
Finalize Business Case summary and recommendation/ rejection	CSI Project Director

Inputs

- CSI Register Of Improvement Projects
- CSI Investment Screening Criteria

Outputs

- CSI Project/Service Costs
- CSI Project/Service Benefits
- CSI Project/Service Timelines
- Business Constraints CSI Project Proposals
- CSI Business Cases
- CSI Register Updates

Analyze CSI Register Projects

This step ranks CSI projects in the register and aligns them to current IT and business issues and needs.

Key Activity	Accountable Role
Assign CSI portfolio ranking	CSI Project Director
Normalize CSI portfolio ranking	CSI Project Director
Review CSI project status and service performance	CSI Project Manager
Identify CSI strategy and portfolio options and scenarios	CSI Project Manager
Model CSI options and scenarios	CSI Project Director
Align the CSI portfolio with key value drivers	CSI Project Coordinator
Model impact of new projects, services and changes on the existing CSI portfolio	CSI Project Coordinator
Define the risks and values for each CSI portfolio initiative in order to prioritize them	CSI Project Coordinator
Prioritize CSI projects and services	CSI Project Director
Update the CSI Register with project rankings	CSI Project Coordinator

Inputs

- Business Case Authorized Recommendation
- Resource Availability
- Approved CSI Improvement projects

Outputs

- CSI Project Recommendations
- CSI Investment Status and Performance
- CSI Register Updates

Authorize CSI Projects

This step authorizes which CSI projects will go forward.

Key Activity	Accountable Role
Review and confirm CSI ranking of projects and investments	CSI Project Director ITSM Steering Committee
Provide direction on each investment to proceed (e.g. cancel, suspend or continue)	CSI Project Director ITSM Steering Committee
Provide funding for investments	ITSM Steering Committee

Inputs

- CSI Register Recommendations
- Resource Availability
- Fit With Current Budgets and Strategies

Outputs

- CSI Register Updates
- Approvals to Proceed
- Authorized Funding

Monitor and Track CSI Projects

This step monitors CSI projects and activities as they execute, reports progress and status and generally steers efforts as they go forward.

Key Activity	Accountable Role
Upload CSI status and performance information	CSI Project Director
Compared expected and realized benefit tracking	CSI Project Director
Monitor impact of new business constraints (new regulation, increase in competition, budget reduction, downsizing, etc.) on planned improvement investments	CSI Project Manager ITSM Steering Committee
Track benefits in accordance with strategic plans	CSI Project Director ITSM Steering Committee
Provide recommendations to CSI plans and strategies as needed	CSI Project Director ITSM Steering Committee
Report on CSI project status	CSI Project Director
Prepare summary of CSI project costs and benefits	CSI Project Manager
Generate and communicate the CSI alignment for the next planning review period	ITSM Steering Committee CSI Project Director

Inputs

- CSI Register information
- CSI Performance and Status information
- Current CSI project costs and revenues
- Planned benefits and actuals realized to date

Outputs

- Updated CSI Register information
- Summary of CSI project costs and benefits
- Recommendations to current CSI plans and strategies based on current portfolio performance and results

Provide CSI Lessons Learned

This step reviews the overall CSI process, control activities and projects undertaken to learn from any issues that may have occurred so they will not be repeated in the future.

Key Activity	Accountable Role
Review current CSI Register status	CSI Project Director
Document and identify CSI issues and concerns	CSI Project Director
Identify future mitigation actions to prevent CSI issues	CSI Project Manager
Document CSI concerns and recommendations as lessons learned	CSI Project Manager

Inputs

- CSI Register information
- Status and Performance of CSI Projects and Services

Outputs

- Lessons Learned

CSI Policies and Controls

It is recommended that CSI activities and the process cycle occur on a regular scheduled basis. This can be as short as 3 months, but should not be any longer than 1 year. The table below presents a typical example of a yearly CSI cycle of activities taken from a number of IT organizations.

CSI Policy	Policy Implications
A formal process exists to investigate and decide which service improvements and projects to provide	• A CSI process owner exists • The process is audited and reviewed annually against its objectives
A model exists to analyze ROI, risk and impact of business changes for each service improvement activity in the CSI Register	• Every service has a documented statement of the investment made in the service • Investments are monitored on a scheduled basis and compared with improvement outcomes and ROI • Each service improvement has documented risks, and counter-measures taken • Customer surveys indicate a high level of satisfaction with the value they are receiving
The CSI Register identifies each service improvement project, the business need, and the improvement outcome to be obtained	• The CSI Register is used as the basis for deciding which service improvement projects to undertake • There is a documented process for defining the improvement benefits and improvement outcome • Each service improvement project in the CSI Register is linked to at least one validated improvement outcome • An audit shows that every service improvement project is documented in the CSI Register
A process exists to review whether service improvements are enabling IT and the business to achieve their strategies	• Feedback is provided on the performance of each service improvement project • An audit shows that improvement outcomes in the CSI Register are consistent with those stated in the strategy
Improvement investments and priorities are updated on a regular basis in response to changes in business need and new technologies where appropriate	• Each CSI Register entry has been evaluated and a decision made about the need for change to the relevant service improvement project • The service improvement projects in the CSI Register align to business strategy and requirements • Customer surveys show continued high levels of satisfaction

CSI Policy	Policy Implications
Capabilities are in place to track service improvement investments throughout the CSI lifecycle	• The investment in each service improvement project is quantified in the CSI Register • Service improvement investments are reported, starting with the initial investment, and followed by monthly, quarterly or annual reporting on ongoing investments • Service improvement investments made are consistent with the projected return on investment forecasts
A formal process exists to evaluate the viability of service improvement projects ending them when they have achieved their objectives or are no longer viable	• Investments are reviewed on a regular basis • Projects may be cancelled or changed based on outcomes and new needs • The CSI Register is updated on a timely basis to provide an accurate status of each service improvement project

CSI Register Example

Construction and management of the CSI Register is somewhat close to that done for a project management portfolio. The register will consist of records. Each record describes one set of activities or a project tied to a distinct service improvement outcome. The kinds of data kept in each record can include:

- Opportunity number

- Name or Short Description

- Description

- Date raised

- Current Status

- Effort (Major, Moderate, Minor)

- Link to Project Plan

- Link to Business Case

- Target Completion Date

- Justification

- Risks if not done

- Success Criteria and Planned Outcomes

- Effort Estimate

- Cost Estimate

- Actual Cost

- Priority Score (based on benefits, costs, risks, alignment with strategies)

- Owner

- Raised By

- To be actioned by

CSI Improvement Cycle Example

It is recommended that CSI activities and the process cycle occur on a regular scheduled basis. This can be as short as 3 months, but should not be any longer than 1 year. The table below presents a typical example of a yearly CSI cycle of activities taken from a number of IT organizations.

Business Cycle	CSI Activities
First Quarter (Jan-Mar)	• Release new CSI plans and activities • Launch new CSI initiatives • CSI Register check and validation • Current and previous CSI cycle register updates • Strategic and tactical input from ITSM and Business Leadership
Second Quarter (Apr-Jun)	• Review and readjust capital allocation for current year • Identify CSI initiatives for next CSI cycle or year • Review current CSI initiatives and pending proposals current status against the ITSM service strategy • Update CSI Register with current CSI initiatives • Conduct strategic look ahead for CSI initiatives
Third Quarter (Jul-Sep)	• Plan and prioritize next year's CSI initiatives • Revise long term capital plans • Confirm CSI project sponsors • Define CSI project outcomes • Pitch CSI plans and strategies to ITSM and Business Leadership • Review current CSI Register and pending proposals against current status and outcomes
Fourth Quarter (Oct-Dec)	• Finalize next year's CSI initiatives • Receive budget proposal for CSI initiatives • Allocate capital to CSI projects • Review current CSI Register and pending proposals against current status and outcomes

Chapter

17

Building a Service Step-By-Step

"Do what you do so well that they will want to see it again and bring their friends!"
—Walt Disney—Founder of Disney

This book has focused on many concepts for building and executing on an ITSM transition effort. What if your focus is on building a new service? This chapter attempts to put many of the ITSM concepts together to do so. It presents a service development blueprint that you might use to design, build and deploy a new IT service.

The blueprint presented here attempts to be fairly comprehensive. There may be many more steps than what might be needed for a particular build effort. The goal has been to provide you with a comprehensive approach that you can tailor for your own needs. Feel free to use it as a reference when a particular service needs to be built or modified.

At its core, the blueprint is somewhat simple on concept. It uses the IT Service Management lifecycle and processes to build the new service (after all, isn't what it was meant for in the first place?). This approach is then combined with many of the implementation approaches described throughout this book. The overall approach is discussed first, followed by an example of its use in building a new monitoring service.

Service Build Blueprint

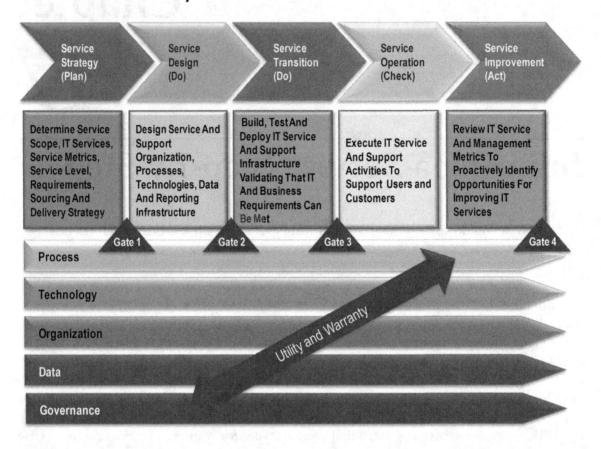

This service building approach will utilize the ITSM service lifecycle activities to identify key strategies for delivering the service, design, build, test, and transition it through support operation. Each of these lifecycle tasks will become a project phase within the approach. At the end of each lifecycle stage, a checkpoint is taken to re-validate intended direction and make sure there is readiness and agreement to proceed to the next stage.

As presented throughout this book before, each stage cuts through five solution project tracks (Process, Technology, Organization, Governance and Project Management) and provides a specific set of work activities and deliverables.

The following pages will describe a more detailed view of how the Service Build Blueprint works to provide a holistic service solution. The detail shows the key activities performed throughout the service build effort.

A high level view of this approach is as follows:

Service Strategy

> In this stage, high level strategies are developed for how the service should be built, operated, delivered and charged for. Key questions to answer at this stage include: "What kind of customers will use this service?" "Who will we partner with to build and

deliver the service?" "How will we market the service?" "How much will we charge for this service?" "How will we organize ourselves to deliver the service?"

Service Design

At this stage, service functions and requirements are translated to design specifications. This includes design of systems involved with managing the service as well as the service itself. A holistic approach of considering people, processes and technologies are designed to deliver and operate the service.

Service Transition

At this stage, the design specifications are translated into a working service solution. The service is built, tested and transitioned to a live production state. Again, a holistic view is taken. People and responsibilities need to be put into place to deliver and operate the service. Processes and procedures need to be implemented and people trained to use them. Supporting technologies to run the service as well as manage it need to be put into place.

Service Operation

In this stage, the service is operated day-to-day. Operational procedures are carried out, incidents logged, requests fulfilled and monitoring of service health and state are examples of general activities carried out to deliver the service to its customers.

Service Improvement

In this stage, service attainment results are reviewed for delivery and quality issues. Proactive activities take place to continually improve the service and ensure it will continue to stay current with the needs of its customers.

The following pages run through each of these stages in more detail.

Service Build Team Organization

When building or making major enhancements to a new or existing service, a service development team should be assembled. The chart below presents a suggested organization for building a new service. The roles shown here are presented in more detail and later mapped into the build activities discussed later.

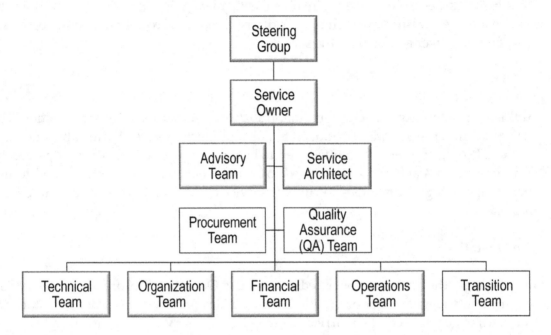

A high level overview of these roles is as follows and discussed within the context of building a service:

Steering Group

> Provides oversight, funding and steering of new service development activities. This group assists with setting direction and vision for the service under development and has final say on issues or conflicts that might arise from development activities

Service Owner

> Owns and leads the development effort end to end. Coordinates all activities involved with building and deploying the service. Optionally, some of the day-to-day build management tasks can be assisted by a project or program manager, but the Service Owner still retains ultimate accountability for the overall build effort.

Advisory Team

> Consist of a mix of business, IT representatives and other service owners that might be dependent on the service being built. This team provides the overall requirements for the new service and validates that those requirements will be met.

Service Architect

Has key accountability for building and architecting the service end to end translating Advisory Team requirements to a workable IT service solution. Provides detailed specification for the service and makes sure that chosen solutions will operate in a functionally correct manner.

Procurement Team

Handles all procurement activities needed to support the service development effort. This can range from managing Request for Proposal (RFP) efforts to coordinating business activities to screen and procure services from outside parties. This team also handles tasks to procure any needed hardware and software that will be needed.

Quality assurance (QA) Team

Accountable to make sure the service product meets business standards for safety and quality. Also validates that the service will meet any legal and regulatory requirements as needed (e.g. SOX or HIPPA as examples).

Technical Team

Accountable for assembling and building the technologies that will underpin the service. This is across the board for applications, servers, network components, service management software or any other technologies that will be needed.

Organization Team

Accountable for all organizational needs to build and support the service such as procuring and training support staff, assigning resources to support roles and managing any communications and organizational change efforts associated with the service. This team may also get involved with marketing types of activities such as developing service brochures, marketing or publishing information about the service being built.

Financial Team

Accountable for building the service cost model, estimating service build and delivery costs, identifying an appropriate service charging approach, identifying requirements for producing customer bills and financial reports. This team also plays a big role in helping to develop the overall business case for the service to executive management.

Operations Team

Accountable for making sure that the service can be operated at acceptable cost and risk to the business. This includes development or reuse of existing operational solutions for handling incidents, problems, changes, monitoring the service, updating any service catalogs and readying support teams and the service desk for live operation of the service.

Transition Team

Has overall accountability for migrating the service from its development state into the live production environment. This includes development of a workable service cutover plan, migration planning, and deployment of the service to a live production state.

The next sections walk through the Service Build approach step by step and map how these roles operate within each build stage.

Service Strategy Building Stage

This stage gets you started by executing a number of work efforts to define the service and how it will be constructed. A step by step list of build tasks is shown below:

Step	Activity	Owning Team	Key Work Product
1	Identify service ownership and steering	Steering Group	Steering Group Membership List Steering Meeting Schedule/Agenda
2	Identify service stakeholders	Steering Group	Stakeholder List
3	Identify service development team	Steering Group	Development Organization Development Team List
4	Analyze markets, customers and needs	Advisory Team	Service Needs Analysis Stakeholder Feedback
5	Develop service perspective, positions, plans and strategies	Advisory Team	Service Change Proposal
6	Identify service value and objectives	Advisory Team	Service Value Description
7	Identify service critical success factors	Advisory Team	List Of Critical Success Factors
8	Develop business demand forecasts that drive service consumption	Advisory Team	List Of Service Demand Drivers Service Business Demand Volumes
9	Develop high level service requirements	Advisory Team	Service Utility Requirements Service Warranty Requirements High Level Functional Requirements Specialized Support Needs Service Targets and Agreements
10	Identify service products and outcomes	Service Architect	Service Product Descriptions
11	Develop service descriptions and features	Service Architect	Service Descriptions Service Features
12	Develop high level service workflows and models	Service Architect	Service Model(s)
13	Define service packaging and options	Service Architect	Service Package Descriptions

Step	Activity	Owning Team	Key Work Product
14	Assess current service delivery capabilities	Service Architect	Support Capabilities Fit
15	Identify key service assets	Service Architect	List Of Needed Service Assets
16	Identify key service constraints	Service Architect	Service Delivery Constraints
17	Identify sourcing strategies and partners	Advisory Team	Hosting Partners Delivery Strategy Service Sourcing Strategy
18	Procure and assign service resources	Service Owner	
19	Develop high level solution architecture	Service Architect	Solution Architecture Standards To Be Used Frameworks To Be Used Identified Reusable Components That Can Be Used By Service
20	Develop high level service build work plans, deliverables and timelines	Service Owner	Service Development Work Plan Service Development Schedule Implementation Timeframes
21	Develop high level test and validation criteria and strategies	Service Architect	User Acceptance Criteria High Level Test Plan Measurable Success Criteria
22	Conduct service business impact assessment	Advisory Team	Business Impact Analysis Results
23	Conduct service risk analysis	Advisory Team	Service Operating Risk and Gaps Service Risk Analysis Results
24	Identify organizational change strategy	Organization Team	Organizational Change Strategy
25	Develop preliminary cost estimates	Financial Team	Service Cost Model
26	Develop service pricing and cost recovery strategy	Financial Team	Service Cost Recovery Strategy Service Contracts
27	Develop documentation and document review strategies	Operations Team	Documentation Repository Documentation Review Procedures

Step	Activity	Owning Team	Key Work Product
28	Identify funding and capital strategies for building the service	Financial Team	Funding Strategy For Service Development Activities
29	Prepare service specification charter	Service Architect	Service Charter Draft
30	Develop entry for new service into service portfolio	Operations Team	Updated Service Portfolio
31	Develop and present business case for building the service	Advisory Team	Agreed Service Business Case
32	Conduct service strategy review gate activities	Service Owner	Go-Forward Approval

Service Design Building Stage

This stage designs the service technologies, functions, processes, organization responsibilities and support model. A step by step list of build tasks is shown below:

Step	Activity	Owning Team	Key Work Product
1	Assemble Design Team, Plans and Schedules	Service Owner	
2	Select Design Approaches and Strategies	Service Architect	Design Standards
3	Procure Service Delivery Suppliers and Partners	Procurement Team	Procurement Requirements
4	Implement Design Knowledge Bases and Tools	Technical Team	Project Repositories
5	Develop Detailed Design Requirements	All Teams	Inventory of Service Design Requirements
6	Design Business and Management Process Workflows	Service Architect	Functional Architecture Process Designs
7	Design Service Data and Information Elements	Service Architect	Data Architecture Data Models Data Designs
8	Design Application Elements	Service Architect	Application Designs Software Package Customizations
9	Design Request Fulfillment Models	Operations Team	Designed Request Work Flows and Models
10	Design Quality Measurements and Indicators	Transition Team	Inventory of Service Metrics
11	Design Service Delivery Agreements	Procurement Team	Service Contracts and Agreements
12	Design Supplier Interfaces	Operations Team	Designed Supplier Handoffs and Interfaces
13	Design Service Delivery Organization	Operations Team	Support Roles/Responsibilities
14	Design Service Technical Assets and Configurations	Technical Team	Component Configurations
15	Design Networking Elements	Technical Team	Network Designs
16	Design Service Availability	Technical Team	Availability Designs

Step	Activity	Owning Team	Key Work Product
17	Design Service Capacity	Technical Team	Capacity Designs
18	Design Security	Security Team	Security Designs
19	Design Management Tooling Elements	Technical Team	Operational Detailed Requirements
20	Design Monitors, Events, Alerts and Thresholds	Operations Team	Operational Detailed Requirements Monitoring Designs
21	Design Physical Site and Environmental Elements	Operations Team	Operational Detailed Requirements Facilities Designs
22	Design Service Operability Elements	Operations Team	Operational Detailed Requirements Operational Control Tasks
23	Design Service Documentation Elements	Organization Team	Service Design Package
24	Design Service Continuity Elements	Operations Team	Service Continuity Designs
25	Design Service Billing and Chargeback	Financial Team	Service Charging and Billing Forms and Reports
26	Design Service Communications and Training Elements	Organization Team	Service Training Plans Service Training Materials
27	Design Service Testing Strategies and Plans	Transition Team	Testing Requirements
28	Design High Level Transition Strategies and Schedules	Transition Team	Deployment Designs/Strategy
29	Develop Service Design Package	Service Architect	Service Design Package
30	Conduct Service Design Review Gate	Service Owner	Go-Forward Agreement

Service Transition Building Stage

This stage builds and deploys the service into the live production environment. A step by step list of build tasks is shown below:

Step	Activity	Owning Team	Key Work Product
1	Assemble Transition Team, Plans and Schedules	Service Owner	
2	Establish Transition Governance	Transition Team	
3	Develop Detailed Transition Plans	Transition Team	Detailed Transition Plans
4	Conduct Transition Risk Assessment	Transition Team	Transition Risk Assessment
5	Establish Service Release Policies	Transition Team	Release Plan and Policy
6	Finalize Service Acceptance Criteria	Service Architect	Service Acceptance Criteria
7	Build Physical Site Infrastructure Elements	Technology Team	Site Surveys Prepared Sites
8	Build Service Application and Utility Elements	Technology Team	Application Code and Installed Packages Migration/Code Libraries
9	Build Service Management Warranty Elements	Operations Team	Configured Operation Tools and Processes
10	Build Service Billing and Chargeback Elements	Financial Team	Billing Forms and Reports
11	Build Service Quality Metrics and Reporting	Transition Team	Service Reports and Metrics
12	Configure Service For Production Operation	Technology Team	Maintenance Procedures Support Call Lists
13	Build Service Continuity Elements	Technology Team	Implemented Service Continuity Hardware, Software, Network and Procedures
14	Implement Service Delivery and Support Organization	Organization Team	Staffed Support Teams
15	Finalize Internal and External Support Agreements	Procurement Team	Finalized Support Agreements and Contracts
16	Document Service Configurations and Knowledge	Service Architect	Updated Architecture Documentation

Step	Activity	Owning Team	Key Work Product
17	Conduct Service Component Testing	Quality Assurance (QA) Team	Test Results
18	Conduct Service System Integration Testing	Quality Assurance (QA) Team	Test Results
19	Conduct Service Availability and Security Testing	Quality Assurance (QA) Team	Test Results
20	Conduct Service Capacity and Performance Testing	Quality Assurance (QA) Team	Test Results
21	Conduct Service Operational Testing	Quality Assurance (QA) Team	Test Results
22	Conduct Service Quality and Compliance Reviews	Quality Assurance (QA) Team	Validation Results
23	Conduct Service User Acceptance Testing	Quality Assurance (QA) Team	User Acceptance
24	Conduct Service and Support Training and Communications	Organization Team	User/Support Training
25	Deploy Service Elements	Transition Team	Rollout Schedules/Volumes
26	Finalize Service Cost Models and Budgets	Financial Team	Service Cost Models
27	Evaluate Service and Known Errors	Transition Team	Validated Architecture
28	Implement Service Desk and Early Life Support	Transition Team	Early Life Support
29	Conduct Service Transition Review Gate	Quality Assurance (QA) Team	Go-Live Agreement
30	Migrate Service To Production State	Transition Team	Live Service

Service Operation Delivery Stage

These are ongoing activities that typically occur as a service is being delivered. The list provided here may be useful as reference during the Service Design and Transition stages to make sure that all operational elements have been considered in the overall service solution. All activities shown here would be typically performed by service operation teams with the support of their suppliers.

Activity	Key Work Product
Monitor Service Events	Communicated Service Alarms And Responses
Process Service Requests	Logged, Classified, And Fulfilled Service Requests
Process Service Incidents	Logged, Classified And Resolved Incidents
Administer Service Access and Security	Managed Ids And Passwords With Controlled Access To IT Services
Manage Service Operations and Control	Executed Run Scripts And Schedules
Conduct Service Problem Management	Logged, Classified And Resolved Problems
	Updated Known Error Logs
Manage Service Physical Sites	Site Maintenance
	Controlled Physical Access To Sites
Provide Service Desk Operations	Working Call Center To Handle User Issues And Communicate Service Issues And Status Back To Users
Provide Service Technical Management	Skilled Resources Necessary To Maintain And Fix The Service
Provide Service Application Management	Working Applications That Underpin The Service
	Applied Application Fixes And Patches
Manage Service Support and Delivery Staff	Service Support Staff In Place
	Support Staff Skills In Place
	Support Staff Appropriately Available To Support The Service
Conduct Ongoing Service Training and Communications	Trained Support Staff
	Trained Users
	Facilities And Materials For Ongoing Training Needs
Conduct Patch Management Activities	Applied Security Patches

Activity	Key Work Product
Maintain Supporting Hardware, Software and Network	Maintained Hardware, Operating Software And Network Network IP Address Management Network Configuration And Routing Updates Security Firewall Configurations
Monitor Internal and External Supplier Quality	Vendor Performance Reports List Of Vendor Issues Regular Vendor Performance Meetings
Manage Service Capacity and Performance	Service Capacity Plans Service Demand Forecasts
Manage Supporting License Agreements	Maintained Hardware And Software License Agreements License Usage Reports
Manage Service Compliance Adherence	Service Audit Findings And Reports (E.G. HIPAA Or SOX Compliance)
Escalate and Communicate Service Issues	Service Broadcasts And Notifications
Maintain Service Configurations and Knowledge	Updated Configuration Items And Knowledge Articles
Manage Service Recovery and Continuity	Maintained Service Continuity Plans Service Continuity Test Results Service Continuity Communications Archived Service Media And Artifacts
Manage Service Supporting Media	Definitive Media Library Stored And Catalogued Media Used To Underpin The Service
Collect Service Operation Quality Metrics	Operation Reports Operation Dashboards And Views Stored Operational Data
Provide Service Quality Reporting	Service Level Attainment Reports Operating Level Attainment Reports
Monitor Customer Satisfaction	Customer Satisfaction Results Customer Satisfaction Surveys

Service Improvement Stage

These are cyclical activities that typically occur to improve the service over time and ensure it continues to provide value to its customers. This activity should be performed on a regular scheduled basis. Typically this might be once every 3-4 months, twice a year or at least once a year.

Step	Activity	Owning Team	Key Work Product
1	Identify Improvement Strategy	Steering Group	Updated CSI Process and Related Activities
2	Define What Needs To Be Measured	Service Architect	Service KPIs and Measurements
3	Identify Current Measurement Capabilities	Service Architect	Assessment of Measurement and Reporting Tools and Methods
4	Finalize Measurement Plans	Service Architect	Measurement, Reporting and Data Collection Strategy
5	Gather Measurement Data	All Teams	Collected Measurement Data
6	Produce Measurement Reports	All Teams	Service Measurement Reports
7	Analyze Data and Information	Service Architect	
8	Target Opportunities For Service Improvement	Service Owner	CSI Register Additions
9	Identify Opportunity Development Costs, Risks, and Benefits	Service Owner	Improvement Business Case
10	Prioritize and Agree Improvement Opportunities	Service Owner	Prioritized CSI Register Activities
11	Develop Service Improvement Plans and Schedules	Service Owner	Service Improvement Plans Updated Timelines and Schedules Updated CSI Register Entries
12	Implement Improvement Opportunities	Service Architect	Implemented Service Improvement Activities
13	Monitor Impacts of Service Improvements	Service Owner	Validated Service Improvement Results
14	Conduct Service Improvement Review Gate	Service Owner	Updated CSI Register Agreed Going Forward Plans

Monitoring Service Build Plan Example

The following pages describe a more detailed view of how our implementation model is used to provide a holistic monitoring solution. The detail shows the key activities performed throughout the implementation effort. Presented here is the overall blueprint for the approach:

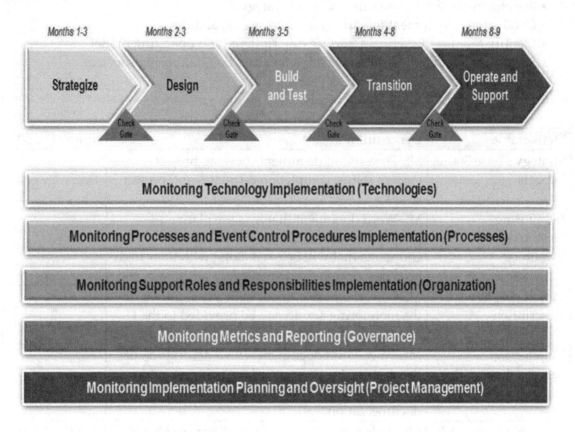

The following pages describe each phase of our planned approach in more detail.

Phase 1: Monitoring Service Strategy

Estimated Duration: 8 Weeks

Key objectives for this phase include:

- Develop the overall Monitoring service strategy
- Conduct the assessment for current monitoring capabilities
- Identify Monitoring services, service descriptions, and operating model
- Identify the detailed roadmap for implementing the Monitoring services

Monitoring Service Strategy Deliverables:

Deliverable	Description	Project Week
Strategy Phase Status Reports	MS-PowerPoint document showing project current status, completion estimates and project issues	Every 2 weeks
Monitoring Technology Inventory	MS-Word document that lists the planned monitoring technologies, interfaces and endpoints that will be used along with their high level descriptions and versions.	2
Monitoring Service Description	MS-PowerPoint document with a description of the monitoring services to be provided to the IT and business units	2
Monitoring Services Knowledge Repository Strategy	MS-PowerPoint document with a description of the strategy for storing and accessing monitoring process, technology and solution as-built documentation	2
Strategy Phase Work Plan	MS-Project document showing Strategy Phase work tasks, milestones and estimated schedules	2
Monitoring Project Kickoff Presentation	MS-PowerPoint document that describes the overall project approach, plans, benefits, scope and monitoring services—this will be a key document for communicating what the project is about to others.	2
Project Change Control Process	MS-Word document that describes the process to be used for requesting, reviewing, escalating, managing and agreeing changes to project scope and activities.	2
Monitoring Services Operating Model	MS-PowerPoint document with a description of the future state monitoring services operating model that will be used to deliver monitoring services	4

Deliverable	Description	Project Week
Monitoring Metrics Strategy	MS-PowerPoint document describing the key metrics, collection and reporting strategy that will be used to evaluate and report success of the implemented monitoring solution	4
Monitoring Service Staging Environment Strategy	MS-Word document that describes the technology and operating requirements for the staging environment used to build the monitoring solution	4
Monitoring Assessment Document	MS-Word document with the Assessment of the existing Monitoring Environment as well as recommendations for improvement of the monitoring high level architecture	4
Monitoring Technology Functional Architecture	MS-PowerPoint document describing the high level technology functional architecture and how the planned toolsets will integrate and fit together	6
Monitoring Service Test Strategy	MS-Word document that describes the high level approach for testing the monitoring solution	6
Monitoring Availability, Backup and Recovery Strategy	MS-PowerPoint document that describes the backup, availability and recovery strategy for the event monitoring tools, data and solution	6
Organizational Change Strategy	MS-PowerPoint document describing planned approach for communicating monitoring progress and services to stakeholders, suppliers and IT units.	6
Monitoring Transition Strategy	MS-PowerPoint document describing the high level transition approach for migrating to the future state monitoring platforms, services, processes and organization	8
Monitoring Detailed Implementation Plan	MS-PowerPoint document showing overall detailed plans and roadmap for the remainder of the monitoring effort and an MS-Project document showing the detailed tasks	8
Monitoring Cost and Risk Summary	MS-Word document that summarizes the implementation, operating risks and high level ongoing operating costs for the monitoring solution	8
Phase 2 Monitoring Service Design Detailed Plan	MS-Project file that has the detailed tasks, milestones and deliverables for the Phase 2 service design effort	8

Monitoring Service Strategy Activities

Process Activities

- Identify Monitoring Services and Users/Customers
- Document Monitoring Service Features
- Document Monitoring Service Delivery Targets and Options
- Assess Current Monitoring Process Capabilities
- Conduct Gap Analysis on Process Capabilities
- Develop Monitoring Process Recommendations
- Define Monitoring Services Operating Model
- Develop high level monitoring process transition plan
- Develop monitoring process detailed implementation plan

Technology Activities

- Inventory current and planned monitoring toolsets
- Review and assess current monitoring implementation activities
- Review and assess current monitoring technologies (policies, customizations, etc.)
- Review and assess current agent deployment (versions, method of deployment and upgrade, etc.)
- Review and assess current maintenance and management of Monitoring infrastructure
- Conduct tooling design workshop
- Determine discovery scope and requirements
- Specify hardware requirements for tooling probe(s)
- Gather IP ranges for discovery
- Gather credentials for foundation discovery
- Gather application specific credentials (web server, database, web servers)
- Conduct gap analysis on monitoring technologies
- Provide Recommendations for Improvement for Monitoring
- Develop Monitoring Technology Functional Architecture
- Develop high level monitoring technology transition plan
- Develop technology detailed implementation plan

Organization Activities

- Identify Key Monitoring Services Stakeholders
- Review and assess current monitoring responsibilities
- Review Supplier Monitoring Capabilities and Services
- Identify and Document Key Monitoring Operating Roles
- Develop Target Monitoring Organization Strategy
- Inventory required monitoring skills and skill levels
- Review current monitoring skills capabilities
- Map Supplier Services to Monitoring Operating Model
- Identify Needed Supplier Delivery Targets
- Conduct gap analysis on monitoring responsibilities and skills

- Provide Recommendations for Monitoring responsibilities and skills
- Develop monitoring Skills Acquisition and Resource Plan
- Develop Organizational Change and Communications Strategy for Monitoring Effort
- Develop high level monitoring organization transition plan
- Develop organizational implementation plan for monitoring organization

Governance Activities

- Build Monitoring Governance Model
- Identify Key Monitoring Governance Roles
- Develop and Manage Project Change Control Process
- Develop Monitoring Critical Success Factors (CSFs)
- Develop Monitoring Key Performance Indicators (KPIs)
- Identify Metrics for Monitoring Service Delivery and Support
- Develop High Level Monitoring Metrics Strategy
- Develop Requirements for Monitoring Solution Service Reporting
- Establish monitoring process and knowledge repository strategy
- Develop and Integrate High Level Monitoring Transition Plan
- Integrate Monitoring Recommendations from Each Sub Team
- Integrate Monitoring Assessment Results and Develop the Assessment Report
- Develop the monitoring solution backup, availability and recovery strategy
- Develop high level monitoring solution test strategy
- Conduct Strategy Phase Checkpoint Gate

Project Management Activities

- Finalize Monitoring Implementation Team
- Finalize Monitoring Strategy Phase Work Plan
- Conduct Project Kickoff Meeting
- Manage The Strategy Phase Work Effort
- Report On Project Status
- Escalate and Communicate Project Issues Where Needed
- Review and track Strategy Phase Deliverables
- Build the Phase 2 Monitoring Service Design detailed work plan
- Monitoring Service Strategy Completion Criteria

The following criteria will be used as part of the checkpoint gate for this phase to indicate its successful completion:

- Timely completion of the Monitoring Service Strategy deliverables
- Confirmation of planned Monitoring Service benefits and costs
- Assessment document includes recommendations for:
- Monitoring (policies, customizations, etc.)
- Agent deployment (versions, method of deployment and upgrade, etc.)
- Maintenance and management recommendation and guidelines
- Planned architecture of monitoring infrastructure.

Phase 2: Monitoring Service Design

Estimated Duration: 12 Weeks

Key objectives for this phase include:

- Design the Monitoring services and support technologies
- Design the Monitoring services and support processes
- Design the Monitoring services and support roles and responsibilities

Monitoring Service Design Deliverables

Deliverable	Description	Project Week
Design Phase Status Reports	MS-PowerPoint document showing project current status, completion estimates and project issues	Every 2 Weeks
Event Design Process	MS-Word document that describes a consistent repeatable methodology and process for requesting, evaluating, designing and maintaining events	2
Monitoring Service Knowledge Repository	Implemented knowledge repository for storing, sharing and communicating monitoring design and implementation artifacts	2
Monitoring Service Maintenance and Improvement Program Design	MS-Word document describing the design for activities to grow and enhance the monitoring solution post project	2
Monitoring Service Staging Environment Design	MS-Word document that describes the detailed technology requirements and operating requirements for the staging environment used to build the monitoring solution	2
Monitoring Service Management Process Design Document	MS-Word document that describes the activities and process flows for event management processes	4
Monitor Tool Component Design Document	MS-Word document, which contains the monitor tool platform design and configuration documentation	4
CMDB Component Integration Design Document	MS-Word document which contains the description of CI types/CIs exported from sources and how these will be imported into the CMDB	4

Deliverable	Description	Project Week
Inventory Of Monitoring Events, Alarms And Thresholds	MS-Word document with inventory of selected monitored events agreed to be within scope of the implementation effort and their escalation paths	4
Monitoring Service Organizational Design Document	MS-Word document describing the detailed operating roles, responsibilities, required skills and skill levels	4
Monitoring Service Training and Communication Design Document	MS-Word document describing the detailed training plan and approach as well as the communications campaign for the monitoring solution	4
Monitoring Dashboard Component Design Document	MS-Word document, which contains the monitoring dashboard design and configuration documentation	6
Monitoring Service Backup, Availability and Recovery Design Document	MS-Word document that describes how the event system and solutions will be backed up, made available and recovered to reduce operating risks in the event of a major business disruption	8
Monitoring Service Security Design Document	MS-Word document that describes how the event system and solutions will be customized to meet established IT security policies and requirements	8
Monitoring Service Metrics and Reporting Design	MS-Word document describing the detailed activities and designs for collecting data and reporting on monitoring activities	8
Monitoring Service Solution Sizing Design	MS-Word document that describes the estimates for solution technology capacity and support labor needed to operate the monitoring solution on an ongoing basis	8
Monitoring Service Test Design	MS-Word document that describes the detailed plans, requirements and activities for testing the monitoring solution	10
Monitoring Service Transition Design	MS-Word document describing the detailed migration plan and activities for moving the event solution to a production state	10
Event Control Procedures	MS-Word document that describes event control procedures for in-scope events	10
Monitoring Service Technical Design Document	Summarizes and integrates the component designs into a comprehensive designed solution for the planned monitoring services	12
Phase 3 Build and Test Phase Detailed Work Plan	MS-Project file that has the detailed tasks, milestones and deliverables for the Phase 3 build and test effort	12

Monitoring Service Design Tasks

Process Activities

- Collect existing process documentation
- Identify and inventory event control procedures
- Develop and document event design process
- Develop and document monitoring event control processes
- Develop monitoring process designs for managing monitoring activities
- Develop monitoring detailed procedures to support monitoring activities
- Identify and design integration points for monitoring processes with other IT and supplier management processes
- Develop monitoring process design document
- Review process designs and monitoring suppliers

Technology Activities

- Review results of tooling and technology assessment
- Design and implement the monitoring knowledge repository
- Design monitor tooling server setup
- Design agent migration steps
- Design agent installation steps
- Design monitor tooling module customization
- Identify and select templates/policies/scripts from existing monitors to be migrated
- Design integration for existing discovery tools and events
- Design health indicators for application events
- Design monitor reports
- Design monitor dashboard views to provide health indicators
- Design events that will be integrated into the monitoring dashboards
- Design CI (Configuration Items) integration with CMDB
- Design procedures for export and import of CIs
- Design application maps
- Design views based on requirements
- Design event correlation for in-scope events
- Create inventory of monitored events, alarms and thresholds
- Design technologies needed to support backup, availability and recovery of the event monitoring solution
- Design monitoring solution staging environment
- Design estimates for event system technology sizing and capacity
- Create Technical Design Document

Organization Activities

- Design event system operating roles and responsibilities
- Document monitoring support roles and tasks
- Identify and document receiving groups and organizations for events
- Design required operating skills and skill levels
- Design estimates for event solution labor and support personnel
- Design event solution training and communications plans

Governance Activities

- Review event designs and integrate designs where needed
- Conduct design reviews with IT and supplier groups
- Coordinate resolution of design issues where needed
- Identify event management security requirements
- Design detailed monitoring transition plans and activities
- Design ongoing monitoring maintenance and improvement program
- Design monitoring metrics collection and reporting
- Design monitoring solution backup, availability and recovery activities
- Design monitoring solution test plans and activities
- Conduct Design Phase Checkpoint Gate

Project Management Activities

- Finalize Monitoring Design Phase Work Plan
- Manage The Design Phase Work Effort
- Report On Project Status
- Escalate and Communicate Project Issues Where Needed
- Review and track Design Phase Deliverables
- Design the Build and Test Phase detailed work plan
- Monitoring Service Design Completion Criteria

The following criteria will be used as part of the checkpoint gate for this phase to indicate its successful completion:

- Timely completion of the Monitoring Service Design deliverables
- Confirmation of planned Monitoring Service benefits and costs

Phase 3: Monitoring Service Build and Test

Estimated Duration: 12 Weeks

Key objectives for this phase include:

- Build the monitoring solution staging environment
- Build the Monitoring services and support solution into the staging environment
- Conduct unit, system and integration tests for the monitoring solution

Monitoring Service Build and Test Deliverables

Deliverable	Description	Project Week
Implemented Monitoring Service Staging Environment	Installed staging environment to support monitoring solution build and test activities	2
Installed Discovery Tool Base Components And Patches	Installed agents and policies on mid-range operating platforms	4
Installed Monitoring Platform And In-Scope Events	Installed monitoring base platform and configurations	6
Installed Specialized Monitors	Installed specialized monitoring base platform and configurations	8
Installed Export/Import Of CI Types From Data Sources	Import and export of selected CI types between monitoring platforms and CMDB	8
Installed Health Indicators	Installed and configured health indicators and events	10
Installed Monitoring Consoles And Dashboards	Implemented and configured monitoring dashboards and views	10
Installed Monitoring Service Metrics And Reports	Installed reporting metrics and reports with delivery of an initial set of management reports	10
Installed Security Customizations	Installed security customizations to meet IT security policies	10
Monitoring Solution Unit, System And Integration Test Results	MS-EXCEL document that inventories each test and the results achieved	12
Updated Monitoring Service Technical Design Document	Updates to the Monitoring Service Technical Design document based on installation and testing of the installed components	12

Deliverable	Description	Project Week
Monitoring Services Training And Communications Guides	MS Office compatible guides to be used for training and communications purposes—these may also include externally produced vendor training and related artifacts all of which are stored into the knowledge repository	12
Phase 4 Transition Phase Detailed Work Plan	MS-Project file that has the detailed tasks, milestones and deliverables for the Phase 4 transition effort	12

Monitoring Service Build and Test Tasks

A summary of the key activities for this phase by project track include:

Process Activities

- Document and store event control procedures
- Assemble and store event management work instructions and process flows
- Update existing support processes and process flows as needed
- Conduct use case testing for event management processes
- Validate technology implementation of event policies and control procedures

Technology Activities

- Install monitoring tool software and patches
- Upgrade monitoring agents to latest version
- Install monitoring agents on midrange servers
- Deploy applicable monitoring and event policies
- Integrate monitoring events and topology with alerting platforms
- Configure discovery tools to monitor availability of Applications
- Unit Test monitoring infrastructure
- Create Health Indicators for Application events
- Create user logons and ids for monitoring infrastructure
- Unit test event recognition
- Unit test Application Health Indicators
- Unit test Business Process Monitors and developed scripts
- Configure Monitoring Reporting and Graphing
- Validate data collection and reports
- Create Dashboard views to provide health indicators
- Integrate events into monitoring consoles
- Create Dashboard views
- Unit test dashboard views and events
- Create exports for CI Types and Data between CMDB and monitoring platforms
- Create integration adapter from data sources to discovery tools
- Verify correct data is exported for CI Types
- Document procedures for export and import of CIs

Organization Activities

- Build and develop Monitoring Service training and communication artifacts

Governance Activities

- Coordinate resolution of event solution build issues where needed
- Update monitoring transition plans and activities based on build activities
- Build ongoing monitoring maintenance and improvement program artifacts
- Build monitoring metrics collection and reporting
- Build monitoring solution backup, availability and recovery activities
- Conduct system and integration tests of the future state monitoring infrastructure
- Document monitoring solution test results
- Review test results with IT and suppliers
- Conduct Build and Test Phase Checkpoint Gate

Project Management Activities

- Finalize Monitoring Build and Test Phase Work Plan
- Manage the Build and Test Phase Work Effort
- Report On Project Status
- Escalate and Communicate Project Issues Where Needed
- Review and track Build and Test Phase Deliverables
- Design the Transition Phase detailed work plan
- Monitoring Service Build and Test Completion Criteria

The following criteria will be used as part of the checkpoint gate for this phase to indicate its successful completion:

- Timely completion of the Monitoring Service Build and Test deliverables
- Successful Monitoring Service Unit, System and Integration Test Results
- Confirmation of planned Monitoring Service benefits and costs

Phase 4: Monitoring Service Transition

Estimated Duration: 12 Weeks

Key objectives for this phase include:

- Train IT and supplier support staff on monitoring solution support and operations
- Conduct user acceptance testing of the monitoring solution
- Validate operational readiness of the monitoring solution
- Transition the monitoring solution to production state

Monitoring Service Transition Deliverables

Deliverable	Description	Project Week
Migrated User Access To Monitoring Infrastructure	User accounts migrated with required access permissions to the monitoring infrastructure	2
Monitoring As-Configured Document	MS-Word document, which contains Monitoring As-Configured Documentation	2
Discovery Tool Infrastructure As-Built Document	MS-Word document, which contains Discovery Tool Infrastructure As-Configured Documentation	4
Monitoring Dashboard As-Built Document	MS-Word document, which contains Monitoring Dashboard As-Configured Documentation	6
CMDB Integration Document	MS-Word Document which contains the description of CI types/CIs exported between monitoring infrastructure and CMDB	6
Policies And Scripts From Monitoring Servers	Templates/policies/scripts from monitoring servers	8
Supplier Agreements For Monitoring Support Responsibilities	Documented agreements from suppliers as needed for support and execution of monitoring services	8
Monitoring Service Trained Support Personnel	Completed training activities for support personnel and demonstrated support capabilities	10
Monitoring Service Operational Readiness Checklist	Completed checklist of operational readiness criteria for solution go-live	12
User Acceptance Test Results	MS-EXCEL document that summarizes the results of the user acceptance testing effort	12

Monitoring Service Transition Tasks

Process Activities

- Support process training efforts
- Assess process capabilities for assigned support resources
- Validate event escalation procedures for correct escalation to appropriate support groups and suppliers
- Assess event control procedures for operational readiness

Technology Activities

- Migrate applicable users from existing monitoring servers
- Migrate applicable templates/policies/scripts from existing servers (based on Monitoring Assessment)
- Deploy new monitoring servers
- Deploy Agents
- Deploy configured discovery events
- Deploy Health Indicators for Application events
- Deploy users
- Deploy monitoring scripts
- Deploy Reporting and Graphing
- Validate deployed diagnostics data collection and reports
- Deploy Dashboard views and health indicators
- Deploy Console events
- Export CI Types and CI Data from monitoring infrastructure
- Verify correct data is exported for CI Types
- Deploy procedures for export and import of CIs
- Finalize as-built documentation for event solution components

Organization Activities

- Assign IT and supplier resources to Monitoring Service roles and responsibilities
- Conduct support training, knowledge transfer and communications activities
- Validate resource capabilities to support monitoring services

Governance Activities

- Establish and finalize supplier agreements for monitoring responsibilities as needed
- Assess operational readiness for monitoring services
- Perform user acceptance testing of monitoring services
- Conduct Transition Phase Checkpoint Gate

Project Management Activities

- Finalize Monitoring Transition Phase Work Plan

- Manage the Transition Phase Work Effort
- Report On Project Status
- Escalate and Communicate Project Issues Where Needed
- Review and track Transition Phase Deliverables
- Monitoring Service Transition Completion Criteria

The following criteria will be used as part of the checkpoint gate for this phase to indicate its successful completion:

- Monitoring solution is installed and operational on IT infrastructure.
- Applicable users have been migrated to new monitoring infrastructure.
- Agents have been deployed and are now reporting to the new monitoring server
- IT infrastructure servers and applications have agents installed and are reporting alerts to the new monitoring server
- Applicable monitoring policies have been migrated to the monitoring servers
- Alerts are integrated with the monitoring infrastructure
- Discovery tools are installed and operational on the IT infrastructure
- Monitoring as-built drawings and documents have been completed
- Event dashboards and consoles are up and running
- Health indicators for Applications have been setup
- Reporting and graphing features are implemented
- The monitoring team is able to validate the data collected and the reports
- Configurations items (CIs) types can be pushed between the CMDB and new monitoring solution
- Export of CIs has been completed and the information has been integrated into the CMDB
- The infrastructure configuration along with the integration procedures is documented.
- Successful Monitoring Service User Acceptance Test Results
- Confirmation of planned Monitoring Service benefits and costs

Phase 5: Monitoring Service Support and Operation

Estimated Duration: 8 Weeks

Key objectives for this phase include:

- Provide support and assistance for the event monitoring solution to IT and supplier support staff
- Establish the going-forward vision for expanding and enhancing the monitoring solution
- Closeout the event monitoring implementation project

Monitoring Service Support and Operation Deliverables

Deliverable	Description	Project Week
Event Solution Support Request Log	MS-EXCEL document showing each support request and its status	Post Project
Event Solution Known Error Log	MS-EXCEL document showing any event solution known errors, their status and work-around information	
Event Solution Improvement Program Vision Document	MS-PowerPoint document showing vision for ongoing event monitoring service expansion and enhancements	

Monitoring Service Support and Operation Tasks

Process Activities

- Establish support processes and procedures
- Provide support assistance for event solution processes

Technology Activities

- Provide support assistance for event technologies

Organization Activities

- Provide support assistance for event roles and responsibilities

Governance Activities

- Log event solution support requests
- Coordinate resolution of event solution support issues as needed
- Document and maintain event solution known-errors and issues
- Document ongoing event management solution improvement and enhancement program

Project Management Activities

- Finalize Monitoring Support plans and resource schedules
- Manage the Support Phase Work Effort
- Report On Project Status
- Escalate and Communicate Project Issues Where Needed

Monitoring Service Support and Operation Completion Criteria

The following criteria will be used as part of the checkpoint gate for this phase to indicate its successful completion:

- Timely completion of the Monitoring Service Support deliverables
- Turnover of project artifacts and deliverables to IT Monitoring Team

About the Author

Randy has over 25 years of extensive hands-on IT Service Management and operations experience gained from many clients around the world. He was the lead author for the ITIL 2011 Service Operation book. He was an early ITIL champion and served a stint as Global Head of Service Management for a worldwide media company with 176 operating centers around the globe. He recently led a major IT Service Management transformation project for a large government agency that included implementation of the latest management technologies as well as process, organizational and governance changes. Randy is also the author of several popular ITIL books: Implementing ITIL, Measuring ITIL, Servicing ITIL and Architecting ITIL. He has also been a frequent speaker around the US for a number of national organizations that focus on IT Service Management. Randy has implemented IT solutions for one client that went on to win a Malcolm Baldrige award for the quality of their IT services. He holds an ITIL V3 Expert and ITIL V2 Service Manager designation and is also Practitioner and ISO20000 Consultant certified. Feel free to contact Randy about any ITSM related concerns, issues or recommended changes and additions to this book. They are always welcome. Randy can be reached at **RandyASteinberg@gmail.com.**